THE OSHA PROCESS SAFETY STANDARD

I0123501

The 30-Year Update

Ian Sutton

Sutton Technical Books

Sutton Technical Books

CONTENTS

PREFACE

I n the year 1992 — 30 years ago — United States Occupational Safety and Health Administration (OSHA) published its Process Safety Management (PSM) regulation (Occupational Safety and Health Administration, 1992). The regulation has not been updated since then. During those 30 years gaps and deficiencies in the regulation have become apparent — often as a consequence of litigation. These gaps require that the standard be updated.

But there are other reasons for revising the standard. During the last three decades the process industries have changed considerably. For example, integrated control systems are now much more prevalent and sophisticated than they were in 1992, and global supply chains have grown and become much more complex. Issues such as these all have an effect on process safety. There is also the increasing pressure on all industries to reduce, or even stop, greenhouse gas emissions. Therefore, the organizations that publish standards, rules and regulation need to ensure that their work reflects this changing world.

OSHA also needs to consider the findings and recommendations of the Chemical Safety Board (CSB). This organization investigates serious accidents in depth and comes up with recommendations for ways of improving process safety. Many of these recommendations call for an upgrade to the process safety regulations.

For all these reasons, in August 2022 OSHA stated that they were opening the standard for modifications and updates. In the agenda for their Stakeholder meeting (Occupational Safety and Health Administration, 2022) the agency identified 24 areas in which they would like to see changes and improvements. The standard was also open for general comments and suggestions.

The fact that OSHA is looking to update the standard does not mean that the process industries have a poor safety performance. Indeed, the good safety record can be attributed in part to the effectiveness of regulations from OSHA and other regulatory bodies. Nevertheless, there is always room for improvement. In the words of the well-known proverb,

> *There is always news about safety, and some of that news will be bad.*

Given this background, we have written this book to provide managers and technical experts in the process industries with a description and analysis of the changes that OSHA is proposing to make.

IMPORTANT NOTE

This book was written after the initial stakeholder meeting and after the comment period was closed, but before the final standard has been issued. Once the final regulation is published this book will be updated, and a second edition will be published.

CHAPTER 1

OSHA'S PROCESS SAFETY MANAGEMENT STANDARD

The 1980s were not a good time for process safety. Among the many incidents that occurred in that decade, two stand out: Bhopal and Piper Alpha. In the year 1984 a release of toxic chemicals from a process plant in the city of Bhopal, India led to the death or serious injury of thousands of people in the local community. Four years later, in 1988, an explosion and fire on the Piper Alpha oil and gas platform in the North Sea resulted in 167 deaths and total loss of the platform. Many more serious incidents in other parts of the world in those years highlighted the problems that the industry faced.

Therefore, as that decade came to a close it was clear that standards and regulations were needed. In response to this concern the Occupational, Safety and Health Administration (OSHA) in the United States issued its process safety management standard in the year 1992 (Occupational Safety and Health Administration, 1992). (The formal title of the regulation is *29 CFR § 1910.119. Process safety management of highly hazardous chemicals.*) The U.S. Environmental Protection Agency (EPA) published its Risk Management Program rule a few years later. The OSHA and EPA standards are intentionally similar to one another.

With respect to offshore oil and gas operations, there was a similar drive toward the implementation of process safety regulations. The Safety Case regime in the U.K. and other European countries was greatly enhanced. In the United States a new agency, the Bureau of Safety and Environmental Enforcement (BSEE), introduced the Safety and Environmental System (SEMS) rule. Finalized in the year 2013, SEMS (Bureau of Safety and Environmental Enforcement, 2022), has many similarities to OSHA's process safety standard.

Since its introduction thirty years ago the OSHA process safety management standard has not been updated or modified. However, the management of process safety has advanced in many ways, and the process industries themselves have changed significantly. Therefore, there was a justification for updating the standard. There was also a need to incorporate findings from incidents that occurred — many of which were investigated in depth by the Chemical Safety Board. Therefore, in August 2022 OSHA proposed an update the standard.

The agenda of their initial Stakeholder Meeting (Occupational Safety and Health Administration, 2022) identified 24 areas for potential change. The standard was also open for comment on all aspects of process safety as can be seen from the preamble of the notice to the Stakeholder Meeting.

> OSHA invites participants to provide public comments related to potential changes to the standard that OSHA is considering.

IMPORTANT NOTE

This book was written after the initial stakeholder meeting and after the comment period was closed, but before the final standard has been issued. Once the final regulation is published this book will be updated, and a second edition will be published.

GOALS FOR THIS BOOK

Given the background just described, this book has the following goals:

1) Briefly describe OSHA's original process safety standard, including an overview of some of the relatively minor changes made since the year 1992 — most of which were the result of legal decisions.
2) Describe the areas where OSHA wants to see changes or improvements in the standard based on the Stakeholder Agenda that was issued in 2022.
3) Discuss other ways in which the process safety standard could be updated and improved.

BOOK STRUCTURE

This book is divided into three main sections. The first section consists of Chapters 1 and 2. Chapter 1 (this one) provides an overview of the current OSHA process safety management regulation. It also provides some background on process safety basics, such as the intent of performance-based regulations. Chapter 2 — Proposed Updates — describes the changes that OSHA plans to make to the standard.

The second section of the book — consisting of Chapters 3 through 26 — provides a short chapter for each of the items that

OSHA listed for potential update. (They are listed in Table 2.2.) Each of these chapters starts with the relevant words of the current regulation. This is followed by OSHA's statements as to what they would like to see changed. There is then a discussion to do with the impact of each proposed change, and how it can be implemented.

The third section of the book, Chapter 27, draws some general conclusions to do with specific changes that OSHA is proposing.

UPDATES

Even after the second edition of the book is published there will be continuing developments and lessons learned as companies implement the new requirements. Therefore, we provide updated information on an on-going basis at the blog https://netzero2050.substack.com/.

THE REGULATION

The purpose of its process safety management standard (Occupational Safety and Health Administration, 1992) is defined by OSHA as follows.

> < The standard > contains requirements for preventing or minimizing the consequences of catastrophic releases of toxic, reactive, flammable, or explosive chemicals. These releases may result in toxic, fire or explosion hazards.

BACKGROUND

The OSHA process safety standard has its roots in the chemical industry. Companies such as Monsanto, ICI Americas, Du-Pont and Dow were instrumental in its development in the late 1980s. These industries generally handle hazardous and corrosive chemicals, often at high temperature and pressure. As the Bhopal catastrophe made clear, an accidental release of these chemicals can have devastating consequences.

The chemical industries possess three important features that affected the content and philosophy of the process safety standard.

First, the chemicals that are used or manufactured are often unique — indeed information about them may be highly confidential and subject to trade secret protection.

Second, the chemical processes in which they are involved are often specialized, complex, and proprietary. This means that there is no way that OSHA can write prescriptive standards for all of these chemicals and industries. This is one of the reasons that the process safety standard is non-prescriptive and performance-based, as discussed in Chapter 2.

A third feature of the chemical industries is that the processes they operate are, by and large, steady-state. Operating conditions do not change much from day to day. Even when conditions do change, the changes are usually gradual (say if a catalyst loses activity). Even batch operations are generally repetitive, not unique. (The steady-state feature of these industries explains why OSHA has had trouble applying its standard to oil and gas-well drilling, as described in Chapter 4. Drilling operations are, by their very nature, constantly changing.)

The process safety standard was later adapted for use in oil refineries and other oil and gas processing operations. Once more, conditions on a typical oil refinery are generally steady-state. Some activities such as the filling and emptying of tanks involve changing conditions, but basically today is much the same as yesterday and tomorrow on these facilities. Moreover, on refineries there is less concern to do with highly hazardous chemicals than on chemical plants (with some exceptions such as the use of hydrogen fluoride in alkylation). However, the potential for fires and explosions is often greater than for chemical plants.

APPLICATION AND SCOPE

Many of the proposed changes to the standard are to do with limits and definitions of process boundaries. Therefore, it is useful to consider the 'Application' section of the standard. It states,

(1) This section applies to the following:

(i) A process which involves a chemical at or above the specified threshold quantities listed in appendix A to this section;

(ii) A process which involves a flammable liquid or gas (as defined in 1910.1200(c) of this part) on site in one location, in a quantity of 10,000 pounds 4535.9 kg) or more except for:

(A) Hydrocarbon fuels used solely for workplace consumption as a fuel (e.g., propane used for comfort heating, gasoline for vehicle refueling), if such fuels are not a part of a process containing another highly hazardous chemical covered by this standard;

(B) Flammable liquids stored in atmospheric tanks or transferred which are kept below their normal boiling point without benefit of chilling or refrigeration.

(2) This section does not apply to:

(i) Retail facilities;
(ii) Oil or gas well drilling or servicing operations; or,
(iii) Normally unoccupied remote facilities.

Paragraph (1) shows that the standard applies to two types of processes — those that handle hazardous chemicals, and those that store or process flammable materials.

Further discussion to do with process safety boundaries is provided in Chapter 27.

CHEMICAL THRESHOLD QUANTITIES

Appendix A of the regulation identifies those chemicals that OSHA considers to be particularly hazardous above a specified threshold quantity.

> *This appendix contains a listing of toxic and reactive highly hazardous chemicals which present a potential for a catastrophic event at or above the threshold quantity.*

Currently there are 137 chemicals in Appendix A. It provides a threshold quantity for each chemical (in pounds). It also provides the CAS (Chemical Abstract Service) number for each chemical.

It is probable that Appendix A will be expanded, partly as a result of Chemical Safety Board recommendations (see Chapter 7).

LOW CONCENTRATIONS

Many of these hazardous chemicals are present in low concentrations when mixed with other, less hazardous chemicals. In order to address this concern, in the year 2016 OSHA established an interim policy regarding concentrations of chemicals that "may retain their hazardous characteristics even at relatively low concentrations". The agency adopted the "1 percent test" that instructed employers to calculate the total weight of a chemical in a process at a concentration of at least 1 percent when no specific concentration or limit for that chemical is defined in Appendix A (Occupational Safety and Health Administration, 2016). The agency states,

> In determining whether a process involves a chemical (whether pure or in a mixture) at or above the specified threshold quantities listed in Appendix A, the employer shall calculate:
>
> (a) the total weight of any chemical in the process at a concentration that meets or exceeds the concentration listed for that chemical in Appendix A, and
>
> (b) with respect to chemicals for which no concentration is specified in Appendix A, the total weight of the chemical in the process at a concentration of one percent or greater. However, the employer need not include the weight of such chemicals in any portion of the process in which the partial pressure of the chemical in the vapor space under handling or storage conditions is less than 10 millimeters of mercury (mm Hg). The employer shall document this partial pressure determination.
>
> In determining the weight of a chemical present in a mixture, only the weight of the chemical itself, exclusive of any solvent, solution, or carrier is counted.

FLAMMABLE MATERIALS

The second type of covered facility is one that handles "10,000 pounds or more of flammable liquids or flammable gases". This 10,000 lb. threshold is quite low. Many industrial and commercial facilities that would not think of themselves as being part of the process industries handle flammable chemicals and fuels that put them well above this threshold. For example, a tank truck hauling gasoline or diesel typically has a load of around 70,000 lb., or nearly 32,000 kg. Many of the difficulties to do with interpreting the standard have been to do with understanding this 10,000 lb. threshold.

PROCESS SAFETY BOUNDARIES

One of the challenges that OSHA faced in 1992, and that it continues to face, was defining the limits of its authority and its relationship with other agencies. For example, tanker trucks (road vehicles) are excluded from the OSHA standard because they are regulated by the Department of Transportation. Oil and gas production facilities have also been excluded, although, as we see in Chapters 4 and 5, they may be included in the future. There are many potential difficulties to be overcome when considering jurisdictional boundaries.

Given this background, many of the proposed updates are to do with defining process safety boundaries with greater precision. In particular, five of the proposed changes to the standard are to do with these boundaries. They are:
- Atmospheric Storage Tanks (Chapter 3),
- Oil and Gas–Well Drilling (Chapter 4),
- Oil and Gas Production Facilities (Chapter 5),
- Retail Facilities Exemption (Chapter 9), and

- Defining the Limits of a PSM–Covered Process (Chapter 10).

DEFINITIONS

Some of the proposed changes to the regulation are to do with definitions of terms. Therefore, the relevant section in the existing regulation is quoted in full in Table 1.1.

Table 1.1. Definitions

Atmospheric tank	means a storage tank which has been designed to operate at pressures from atmospheric through 0.5 p.s.i.g. (pounds per square inch gauge, 3.45 Kpa).
Boiling point	means the boiling point of a liquid at a pressure of 14.7 pounds per square inch absolute (p.s.i.a.) (760 mm.). For the purposes of this section, where an accurate boiling point is unavailable for the material in question, or for mixtures which do not have a constant boiling point, the 10 percent point of a distillation performed in accordance with the Standard Method of Test for Distillation of Petroleum Products, ASTM D–86–62, which is incorporated by reference as specified in §1910.6, may be used as the boiling point of the liquid.
Catastrophic release	means a major uncontrolled emission, fire, or explosion, involving one or more highly hazardous chemicals, that presents serious danger to employees in the workplace.
Facility	means the buildings, containers or equipment which contain a process.

Highly hazardous chemical	means a substance possessing toxic, reactive, flammable, or explosive properties and specified by paragraph (a)(1) of this section.
Hot work	means work involving electric or gas welding, cutting, brazing, or similar flame or spark-producing operations.
Normally unoccupied remote facility	means a facility which is operated, maintained or serviced by employees who visit the facility only periodically to check its operation and to perform necessary operating or maintenance tasks. No employees are permanently stationed at the facility. Facilities meeting this definition are not contiguous with, and must be geographically remote from all other buildings, processes or persons.
Process	means any activity involving a highly hazardous chemical including any use, storage, manufacturing, handling, or the on-site movement of such chemicals, or combination of these activities. For purposes of this definition, any group of vessels which are interconnected and separate vessels which are located such that a highly hazardous chemical could be involved in a potential release shall be considered a single process.

Replacement in kind	means a replacement which satisfies the design specification.
Trade secret	means any confidential formula, pattern, process, device, information or compilation of information that is used in an employer's business, and that gives the employer an opportunity to obtain an advantage over competitors who do not know or use it. Appendix D contained in §1910.1200 sets out the criteria to be used in evaluating trade secrets.

MANAGEMENT PRINCIPLES

The process safety standard differs from most other safety regula-tions because it is performance-based, non-prescriptive, and often calls for the use of risk analysis.

PERFORMANCE-BASED

Process safety regulations provide very little specific detail as to what has to be done. Basically, they say, "Do whatever it takes on *your* facility not to have accidents. How you do that is your call". It is up to the managers, technical advisers and the operations/ maintenance personnel to determine what needs to be done. The only measure of success is success.

NON–PRESCRIPTIVE

Performance-based regulations are necessarily non–prescriptive. They do not tell company managers what to do, they merely say "Be safe". As we have already seen, one reason for taking this approach is that there are thousands of chemical plants, many of which use proprietary process technology. There is no way that OSHA, or any other agency, can write prescriptive rules for all of these facilities, or for potentially unsafe situations. Moreover, information to do with chemicals and process technology is often protected by trade secrets.

This lack of detail helps explain why the OSHA PSM standard is so short — it simply requires that a program be in place, and that it works. The regulation consists of eight pages of text, supplemented with Appendices and Examples.

RISK ANALYSIS

Because non-prescriptive regulations do not provide specific rules as to what is acceptable, they have to be risk-based. But risk can never be zero — hazards are always present, those hazards can have serious consequences, and there is always a chance that a hazard will move from being a potential event to an actual incident. Therefore, managers have to devise means for measuring risk as objectively as possible. They also have to determine the value for "acceptable risk" — something that is very difficult to do because risk is, fundamentally, a subjective matter.

The fact that risk can never be zero also means that process safety programs are never complete — there is always room for improvement. No company, no matter how good its process safety program, is perfect. There is always a chance of something going badly awry.

COMPLIANCE

A risk-based approach to safety means that, from a theoretical point of view, it is never possible to be "in compliance". No matter how well run a facility may be, accidents will eventually occur because risk can never be zero. Hence compliance means that a company aims to meet the spirit of the regulation, not merely the letter of the law.

OSHA has not included guidance to do with risk analysis or risk assessment in its proposed updates to the standard. A possible reason for this decision is that no regulatory body wants to be involved in the vexed topic of acceptable risk. If they define a boundary below which a certain level of risk is acceptable, then they have implicitly said that it is acceptable to have a catastrophic accident and that fatalities will occur. No government agency wants to be involved in that type of discussion.

OCCUPATIONAL SAFETY

It is important to distinguish between process safety and occupational safety. Process safety is primarily concerned with process-oriented issues such as uncontrolled chemical reactions, equipment corrosion, and the inadvertent mixing of hazardous chemicals. The impact of such events can lead to major incidents such as explosions, large fires and the release of toxic materials. Occupational safety, sometimes referred to as "hard-hat" safety, covers topics such as vessel entry, vehicle movement, protective clothing and tripping hazards. Whereas process safety is mostly to do with system failures, occupational safety is more concerned with the actions and behaviors of individuals and small groups of people.

A good process safety program will likely lead to better occupational safety performance because management systems overall will benefit. However, improvements in occupational safety are less likely to lead to corresponding improvements in process safety.

MANAGEMENT ELEMENTS

At the heart of the OSHA's standard are fourteen management elements listed in Table 1.2.

Table 1.2. The Management Elements

1.	Employee Participation
2.	Process Safety Information
3.	Process Hazards Analysis
4.	Operating Procedures
5.	Training
6.	Contractors
7.	Prestartup Safety Review
8.	Mechanical Integrity
9.	Hot Work Permit
10.	Management of Change
11.	Incident Investigation
12.	Emergency Planning and Response
13.	Compliance Audits
14.	Trade Secrets

Since the year 1992 many organizations and companies have developed alternative, more effective, systems. However, OSHA is not proposing to change this structure as a part of the proposed updates.

CHAPTER 2

PROPOSED UPDATES

IMPORTANT NOTE (repeated from the previous chapter)

This book was written after the initial stakeholder meeting and after the comment period was closed, but before the final standard has been issued. Once the final regulation is published this book will be updated, and a second edition will be published.

OSHA'S RULEMAKING PROCESS

The previous chapter provided an overview of OSHA's current process safety management (PSM) standard. In this chapter we look at some of the changes that the agency is proposing.

The full rulemaking process is divided into the following seven steps:
1. Making the Decision: Conducting Preliminary Rulemaking Activities.
2. Developing the Proposed Rule.
3. Publishing the Proposed Rule.
4. Developing and Analyzing the Rulemaking Record.
5. Developing the Final Rule.
6. Publishing the Final Rule.

7.   Post–Promulgation Activities.

There is no explicit step for updating a rule. However, the material discussed in this book seems to be at Step 4.

EXECUTIVE ORDER 13990

The material in this book focuses on the technical and management aspects of OSHA's process safety management standard. But, before looking at the proposed changes, it is important to be aware of Executive Order 13990 that was signed by President Biden in January 2021. This order provides the background to changes in many environmental and safety regulations, including those from OSHA and the EPA to do with process safety.

The introduction to the order reads as follows,

> *Executive Order 13990, of January 20, 2021, directs Federal agencies to immediately review, and take action to address, Federal regulations promulgated and other actions taken during the last 4 years that conflict with national objectives to improve public health and the environment; ensure access to clean air and water; limit exposure to dangerous chemicals and pesticides; hold polluters accountable, including those who disproportionately harm communities of color and low-income communities; reduce greenhouse gas emissions; bolster resilience to the impacts of climate change; restore and expand our national treasures and monuments; and prioritize both environmental justice and employment.*

This executive order replaced previous orders, some of which had been put in place by the previous administration. These included: EO 13834, EO 13778, EO 13783 and 13807.

THE STAKEHOLDER AGENDA

The agenda for the October 2022 stakeholder meeting (Occupational Safety and Health Administration, 2022) provided the information to do with the proposed changes.

There are 24 proposed changes divided into the following three sections:

I.   Background
II.   Stakeholder Meeting
III. Submitting and Accessing Comments

I. BACKGROUND

OSHA starts the stakeholder document as follows. (Some administrative material has been removed, and phrases that are particularly important to the discussions in this book have been highlighted).

> OSHA published the PSM standard, 29 CFR 1910.119, in 1992 in response to several catastrophic **chemical-release incidents** that occurred worldwide. The PSM standard requires employers to implement safety programs that identify, evaluate, and control highly hazardous chemicals. Unlike some of OSHA's standards, which prescribe precisely what employers must do to comply, the PSM standard is

"performance-based," and outlines 14 management system elements for controlling highly hazardous chemicals. *Under the standard, employers have the flexibility to tailor their PSM programs to the unique conditions at their facilities.*

. . .

Since its publication in 1992, the PSM standard has not been updated. The 2013 **ammonium nitrate explosion** *at a fertilizer storage facility in West, Texas renewed interest in PSM. In response to this incident, on August 1, 2013, Executive Order (E.O.) 13650, Improving Chemical Facility Safety and Security, was signed. The E.O. directed OSHA and several other federal agencies to, among other things, modernize policies, regulations, and standards to enhance safety and security in chemical facilities by completing certain tasks, including: coordinating with stakeholders to develop a plan for implementing improvements to chemical risk managements practices, developing proposals to improve the safe and secure storage handling and sale of ammonium nitrate, and reviewing the PSM and* **Risk Management Plan (RMP)** *rules to determine if their* **covered hazardous chemical lists** *should be expanded.*

. . .

OSHA is holding this stakeholder meeting to reengage **stakeholders** *and solicit comments on the modernization topics mentioned in the RFI and SBAR panel report, as well as any* **additional PSM-related** *issues* stakeholders would like to raise. The list of modernization topics is listed below in Section II.

The words and phrases that are highlighted are discussed below.

- The standard has its roots in "chemical-release" incidents. But, as discussed in the "Application" section of the previous chapter, many facilities that would not regard themselves as being part of the process industries handle hazardous chemicals or flammable/explosive materials. One of the justifications for the proposed updates is to try and define process safety boundaries more precisely.
- As we saw in the previous chapter, the "performance-based" aspect of the standard is fundamental to the very nature of process safety.
- The ammonium nitrate explosion is one of many incidents analyzed and described by the Chemical Safety Board (CSB). The reports from this organization, which is described below, have frequently called on OSHA to update the PSM regulation.
- OSHA recognizes that the EPA is also updating its Risk Management Program (RMP), as discussed later in this chapter.
- OSHA has been directed to coordinate with "stakeholders". The meaning of the word stakeholder is not defined, but it presumably includes everyone who works at a facility, the people living nearby, and stockholders.
- OSHA is proposing to add new materials to the "hazardous chemical lists". (These are listed in Appendix A of the standard.)
- The phrase "as well as any additional PSM-related issues" makes it clear that comments and suggestions are

not limited to the list of topics that OSHA has prepared.

II. STAKEHOLDER MEETING ITEMS

Proposed changes to the document are divided into two groups which together contain 24 proposed updates. The first group is to the scope of the PSM standard itself. It consists of the eight items shown in Table 2.1.

Table 2.1. Proposed Changes to the Standard

1.	Clarifying the exemption for atmospheric storage tanks;
2.	Expanding the scope to include oil-and gas-well drilling and servicing;
3.	Resuming enforcement for oil and gas production facilities;
4.	Expanding PSM coverage and requirements for reactive chemical hazards;
5.	Updating and expanding the list of highly hazardous chemicals in Appendix A;
6.	Amending paragraph (k) of the Explosives and Blasting Agents Standard (§ 1910.109) to extend PSM requirements to cover dismantling and disposal of explosives and pyrotechnics;
7.	Clarifying the scope of the retail facilities exemption; and
8.	Defining the limits of a PSM-covered process.

The second group contains the sixteen items referred to as 'Particular Provisions'. They are shown in Table 2.1. They are based on the fourteen management elements shown in Table 1.2.

Table 2.2. Particular Provisions

1.	Amending paragraph (b) to include a definition of RAGAGEP;
2.	Amending paragraph (b) to include a definition of critical equipment;
3.	Expanding paragraph (c) to strengthen employee participation and include stop work authority;
4.	Amending paragraph (d) to require evaluation of updates to applicable recognized and generally accepted as good engineering practices (RAGAGEP);
5.	Amending paragraph (d) to require continuous updating of collected information;
6.	Amending paragraph (e) to require formal resolution of Process Hazard Analysis team recommendations that are not utilized;
7.	Expanding paragraph (e) by requiring safer technology and alternatives analysis;
8.	Clarifying paragraph (e) to require consideration of natural disasters and extreme temperatures in their PSM programs, in response to E.O. 13990;
9.	Expanding paragraph (j) to cover the mechanical integrity of any critical equipment;
10.	Clarifying paragraph (j) to better explain "equipment deficiencies";

11.   Clarifying that paragraph (l) covers organizational changes;

12.   Amending paragraph (m) to require root cause analysis;

13.   Revising paragraph (n) to require coordination of emergency planning with local emergency-response authorities;

14.   Amending paragraph (o) to require third-party compliance audits;

15.   Including requirements for employers to develop a system for periodic review of and necessary revisions to their PSM management systems (previously referred to as "Evaluation and Corrective Action"); and

16.   Requiring the development of written procedures for all elements specified in the standard, and to identify records required by the standard along with a records retention policy (previously referred to as "Written PSM Management Systems").

Combining Tables 2.1 and 2.2 it can be seen that there is a total of 24 proposed changes.

III. SUBMITTING AND ACCESSING COMMENTS

As with any other proposed regulation, the public had an opportunity to provide their comments on the proposed changes, and what they would like to see. The comment period is now closed, but the comments can be reviewed at Docket No. OSHA–2013–0020 (Occupational Safety and Health Administration, 2022).

SUTTON COMMENT

Your author submitted a comment to OSHA on October 6th 2022 (Sutton, Comment Submitted to OSHA, 2022). It reads as follows,

In the notice to do with the Stakeholder Meeting (August 30, 2022), OSHA included the following paragraph,

8. Clarifying paragraph (e) to require consideration of natural disasters and extreme temperatures in their PSM programs, in response to E.O. 13990;

Issues such as rising water levels and increasing temperatures will also affect the safety of process facilities, and will increase the likelihood of a release of highly hazardous chemicals.

The United States Securities and Exchange Commission (SEC) published a proposed rule in March 2022. Its title is 'The Enhancement and Standardization of Climate-Related Disclosures for Investors' (https://www.sec.gov/rules/proposed/2022/33-11042.pdf). Their proposed rule calls on public companies to:

- *Report their greenhouse gas emissions;*
- *Describe their programs to reduce the greenhouse gas emissions; and*
- *Evaluate the risk that climate change poses to their financial situation.*

All three of these topics are relevant to the process industries and to the process safety regulation.

The process safety "way of thinking" is helpful when considering climate issues. For example, when it comes to hazards analysis the process safety community ,

1. *Identifies the hazards;*
2. *Assesses the likelihood and consequences of those hazards;*
3. *Risk ranks each of the identified hazards; and*
4. *Evaluates the feasibility of proposed responses to control that hazard.*

This approach to identifying and controlling high-risk situations can be applied to climate change problems because they possess many of the same features.

A further area of commonality is to do with systems analysis. Process incidents are rarely simple. They occur within complex systems, and are often difficult to identify or analyze. This justifies the paragraph to do with root cause analysis.

12. Amending paragraph (m) to require root cause analysis;

Similarly, climate change challenges are part of complex systems involving many factors such as resource depletion and economics. Root cause analysis techniques can be helpful in understanding how to effectively address climate issues.

*To conclude: it is suggested that paragraph (e) be
expanded to show how process safety techniques
can help address climate concerns.*

ADDITIONAL FEEDBACK

In addition to the 24 items just descrbed, OSHA looked for feedback on any other issues of interest to their stakeholders. The
following is from the meeting agenda,

*The meeting will feature a brief presentation from
OSHA on the background of the PSM standard and
some of the issues outlined in this notice. After
the presentation, there will be time for registered
commenters to provide verbal comments. PSM
rulemaking topics are outlined in the lists below,* **but
commenters may provide feedback on additional
PSM-related issues.**

‹ my emphasis ›

There was a minimal e response from the process safety community to use this opportunity to rethink the process safety
standard.

THE EPA'S RISK MANAGEMENT PROGRAM

The OSHA regulation obtains its regulatory authority from the
Amendments to the Clean Air Act, signed into law by President H.W.
Bush in the year 1990. That Act also required the EPA (Environmental Protection Agency) to develop and implement process safety
standards. OSHA, as its name implies, is concerned with *occupa-*

tional safety. Hence the agency focuses on the safety and health of workers "inside the fence". The EPA is more concerned with the *environmental* impact of "toxic, fire and explosion hazards" on the general public. In other words, their focus is on events that take place "outside the fence". There is, of course, substantial overlap.

The original authorizing legislation required OSHA and the EPA to develop standards that are similar to one another. This goal was partially achieved, but there are substantial differences between them. One of the biggest differences is that the EPA requires companies to develop a Risk Management Program (RMP) that is submitted to the agency. OSHA does not have such a requirement.

The following is what OSHA says about the EPA program.

> *In the Clean Air Act Amendments of 1990, Congress required OSHA to adopt the PSM standard to protect workers and required EPA to protect the community and environment by issuing the RMP rule. The PSM and RMP rules were written to complement each other in accomplishing these Congressional goals.*

The EPA is developing updates to their Risk Management Program (RMP) standard in parallel with OSHA (Environmental Protection Agency, 2022). The title of these updates is 'Safer Communities by Chemical Accident Preventions'. Many of the comments to do with the OSHA standard will apply to the RMP and *vice versa*. However, the EPA typically provides more detail and guidance than OSHA.

CHEMICAL SAFETY BOARD

One of the reasons that OSHA is re-opening its standard is to accommodate the many recommendations that they have received from the Chemical Safety Board (CSB) — an independent, nonregulatory federal agency that investigates the root causes of major chemical incidents. The CSB was created under the Clean Air Act Amendments of 1990 — the same legislation that authorized both OSHA's process safety program, and the EPA's Risk Management Program (RMP).

The following quotation is from a CSB document, *Drivers of Critical Chemical Safety Change* (Chemical Safety Board, 2018).

> *Both OSHA and EPA have safety management regulations. PSM regulations in the U.S. have undergone little reform since their inception in the 1990s. While there have been some initial steps toward improvements in PSM at the Federal level, a more comprehensive PSM system is needed to protect worker safety, public health and the environment.*

> *Over the last two decades, the CSB has conducted several investigations that identified a need for improvements and modernization to OSHA's PSM and to EPA's Risk Management Plan (RMP) Program. The CSB has noted in its investigations of major refinery incidents that both PSM and RMP, although written as performance-based regulations, appear to function primarily as prescriptive activity-based regulatory schemes that require extensive rulemaking to modify, resulting in stagnation despite the availability of revised*

best practices and technology.

Specifically, CSB investigations of the Tesoro Anacortes refinery explosion and fire in April 2010 and the Chevron Richmond refinery fire in August 2012 found that there was no requirement to reduce risks to As Low As Reasonably Practicable (ALARP). There was no mechanism to ensure continuous safety improvement; no requirement to implement inherently safer technology or the hierarchy of controls; no increased role for workers and worker representatives in process safety management; and there neede to be a more proactive, technically qualified regulator in place.

As a result of these findings, the CSB made recommendations at the Federal, state, and local level to prevent major incidents by adopting a more rigorous regulatory system that requires covered facilities to continuously reduce major hazard risks.

To improve OSHA's PSM Standard and EPA's RMP Program, the CSB has made the following recommendations:

- *Update existing Process Hazard Analysis (PHA) requirements to include the documented use of inherently safer systems, hierarchy of controls, damage mechanism hazard reviews, and sufficient and adequate safeguards;*
- *Develop more explicit requirements for facility/ process siting and human factors, including fatigue;*
- *Add safety-critical equipment to existing*

mechanical integrity requirements; and

- *Require coordination of covered facility emergency plans with local emergency response authorities.*

In addition to the Anacortes incident already reference, CSB reports that have called for OSHA to update their regulation include the following:

- Improving Reactive Hazard Management,
- Synthron Chemical Explosion,
- BP Amoco Thermal Decomposition Incident,
- West Fertilizer Explosion and Fire,
- Reactive Hazards Management, and
- Dangerously Close: Explosion in West, TX.

As well as its written reports, the CSB has developed many highly instructive videos that describe and explain the incidents that they investigated. For example, the video 'Silent Killer: Hydrogen Sulfide Release in Odessa, Texas' illustrates an incident that resulted in two fatalities (Chemical Safety Board, 2021).

OFFSHORE SAFETY MANAGEMENT

As we saw in Chapter 1, the offshore industry has had its own serious challenges when it comes to major incidents. The Piper Alpha catastrophe in the year 1988 was that industry's equivalent of the Bhopal disaster. It demonstrated that a fundamental upgrade to process safety standards and application was needed.

With respect to process safety there are considerable differences between the offshore and onshore industries. For example,

- In an emergency, the personnel at a chemical plant or refinery can usually evacuate from the unit to a safe location. Such is not the case offshore — if a safe secondary location is not available then there is nowhere to go but overboard.
- Offshore platforms are generally congested, it is difficult to achieve adequate spacing between equipment items. Chemical plants are often congested, but not usually to the same degree as an offshore platform.
- Offshore facilities handle toxic chemicals, but usually in small quantities. Moreover, those chemicals are not usually, by chemical plant standards, all that hazardous.
- Offshore facilities do not operate chemical processes at very high temperatures and pressures. Nor do offshore facilities generally need to consider the use of unusual materials of construction.

In spite of these differences, the two industries have influenced one another to a large degree. For example, topics such as Stop Work Authority, which are very important offshore, are now being considered for the OSHA standard.

SAFETY CASES

In most areas of the world, offshore safety is handled through the use of Safety Cases — the major exception being the United States. The Safety Case philosophy is similar to that of the OSHA standard in that it is fundamentally prescriptive and performance based. A company prepares a "case" to show that it is safe. The company then takes the actions needed to achieve that level of safety.

Following the Piper Alpha catastrophe, the offshore industry greatly enhanced the scope and quality of the Safety Case approach. The defining report was published by a team led by Lord Cullen, a Scottish High Court judge (Cullen, 1990).

SEMS

In the year 2010 the offshore drilling rig Deepwater Horizon was destroyed in an explosion and fire caused by a well blowout. Eleven men died, and large quantities of oil were spilled into the Gulf of Mexico. This event led to the creation of a new United States regulatory agency — the Bureau of Safety and Environmental Enforcement (BSEE). This agency issued its own offshore safety standard called the Safety and Environmental Management Systems (SEMS) rule. There is a high degree of overlap between OSHA's PSM regulation and SEMS (Sutton, Offshore Safety Management).

Like OSHA's PSM, the SEMS rule is built around the management elements shown in Table 2.3. It can be seen that there are strong similarities between the SEMS and OSHA management systems that were listed in Table 1.3.

Table 2.3. SEMS Management Elements

1.	General
2.	Safety and Environmental Information
3.	Hazards Analysis / Job Safety Analysis
4.	Management of Change
5.	Operating Procedures
6.	Safe Work Practices
7.	Training

8.   Assurance of Quality and Mechanical Integrity of Equipment

9.   Pre-Startup Review

10.   Emergency Response and Control

11.   Investigation of Incidents

12.   Audit of Safety and Environmental Management Program Elements

13.   Records and Documentation

14.   Stop Work Authority

15.   Ultimate Work Authority

16.   Employee Participation

17.   Reporting of Unsafe Conditions

Three of the SEMS elements are not explicitly found in the OSHA standard. They are Stop Work Authority, Ultimate Work Authority, and Reporting of Unsafe Conditions. OSHA has adopted the Stop Work concept into its updated proposals (the third item in Table 2.2). However, as we discuss in Chapters 4 and 5, there may be a misunderstanding as to how that concept can be applied to chemical plants and other onshore facilities.

CHAPTER 3

ATMOSPHERIC STORAGE TANKS

The first item on OSHA's list of proposed updates is 'Atmospheric Storage Tanks'. This seems to be a rather strange choice, given that this topic is a narrow one, especially when compared with other topics such as Safer Technology and Stop Work Authority. The reason for this being selected first may be that the Meer Case (discussed below) was an early and significant challenges to the OSHA standard, and the issues that it raised have not yet been fully resolved.

THE REGULATION

The following are the elements of the current process safety standard. The topic of this chapter is to do with the first of these: *Application*.

(a) *Application*
(b) Definitions
(c) Employee Participation

(d)   Process Safety Information
(e)   Process Hazards Analysis
(f)   Operating Procedures
(g)   Training
(h)   Contractors
(i)   Prestartup Safety Review
(j)   Mechanical Integrity
(k)   Hot Work
(l)   Management of Change
(m)   Incident Investigation
(n)   Emergency Planning and Response
(o)   Compliance Audits
(p)   Trade Secrets

There are three places in the current standard that refer specifically to atmospheric storage tanks.

Under *Purpose,* paragraph (1)(ii) reads,

> *(B) Flammable liquids stored in atmospheric tanks or transferred which are kept below their normal boiling point without benefit of chilling or refrigeration.*

In paragraph (b) — the Definitions section of the standard — OSHA says,

> *Atmospheric tank means a storage tank which has been designed to operate at pressures from atmospheric through 0.5 p.s.i.g. (pounds per square inch gauge, 3.45 Kpa).*

This definition provides the differentiation between 'atmospheric storage tank' and 'pressure vessel'. The pressure above the liquid space this type of storage tank is either at atmospheric pressure or just above it. As the tank is filled, vapors are discharged through a vent line, either directly to the atmosphere or to a flare system. As the tank is emptied air or an inert gas must be added at a sufficient rate to ensure that the pressure in the tank does not fall below atmospheric, otherwise it may implode.

In paragraph (j) — Mechanical Integrity — OSHA says,

> Paragraphs (j)(2) through (j)(6) of this section apply to the following process equipment: (i) Pressure vessels and storage tanks

These three citations tell us that OSHA treats storage tanks as being part of the normal equipment on a chemical plants or refinery; it does not give them special treatment.

PROPOSED UPDATE

The proposed change is,

> Clarifying the exemption for atmospheric storage tanks

This exemption is not part of the original standard. Instead, it is to do with the 'Meer Decision' that is discussed below.

DISCUSSION

A theme that runs through many of the proposed updates to the standard is the topic of limits. Process facilities have many connections to the 'outside world'. These connections include piping, instrument signals (in both directions) and information transfer. Defining the boundaries is both difficult and important. These difficulties became apparent not long after the process safety standard was published with respect to atmospheric storage tanks.

THE MEER DECISION

In 1995, the Meer Corp. in New Jersey faced numerous citations under the PSM standard. The company challenged some of these citations on the basis that OSHA was overreaching when applying the PSM standard to its atmospheric storage tanks (Secretary of Labor v. Meer Corp., OSHRC Docket No. 95-0341). There is no need to delve into the details of this suit except to recognize that clarification was needed.

> *An administrative law judge ruled that PSM coverage does not extend to flammables stored in atmospheric tanks, even if the tanks are connected to a process. As a result, employers can exclude the amount of flammable liquid contained in an atmospheric storage tank, or in transfer to or from storage, from the quantity contained in the process when determining whether a process meets the 10,000-pound threshold quantity. On May 12, 1997, OSHA issued a Regional Administrator's memorandum acknowledging this decision.*

MEANING OF THE WORD "PROCESS"

The Meer decision still leaves the term "tanks connected to a process" as requiring clarification. For example, when most people who work on a chemical plant or refinery hear the phrase "storage tank" they probably visualize the large API tanks that contain liquids that are fed to a process, or that store liquid products. Generally, these tanks are on the physical perimeter of the process. However, an intermediate or day tank between two parts of the process is also "an atmospheric storage tank". What is the different between a storage tank, a day tank and an intermediate tank?

Regardless of where a storage tank is located, it is always connected to the process because liquids flow into and out of it. Therefore, any atmospheric storage tank could be involved in a process-related hazard. For example, the wrong liquid could be directed into a tank resulting in a serious incident. Or liquid may reverse flow out of a tank into an area where the operating personnel are not trained in how to handle that material. Or, if the vapors from the tank are vented to a flare header, it may be possible for other materials in the header to enter the tank's vapor space.

Looked at this way, tanks are always part of a process. But, in principle, this line of thinking could be extended almost indefinitely. For example, if the tank is padded with nitrogen to keep the gas space inert, then it is possible for process vapors to enter the nitrogen system from the tank space were the nitrogen header pressure to fall. There have to be boundaries somewhere.

NFPA 30

Further analysis and discussion to do with the inclusion of storage tanks in the standard was provided by the National Fire Protection Association (NFPA) in a formal comment from Greg Cade, Division Director, Government Affairs. The essence of their response is that there is no need for an extension to the standard to cover atmospheric storage tanks. The letter states that companies are already required to follow another OSHA standard (§1910.106) which references NFPA 30.

> *OSHA already covers atmospheric aboveground storage tanks under Subpart L, §1910.106 (see item 4 below), which is, in turn, based on NFPA 30-1969. Since the provisions for atmospheric aboveground storage tanks in NFPA 30 have not changed appreciably since the standard was issued in 1969, §1910.106 actually provides an appropriate degree of governance over these tanks. To bring these tanks under PSM would duplicate regulations OSHA already has in place, with no discernable benefits and with the possibility of future conflicts.*

Details to do with sections 21 and 22 of NFPA follow.

The letter concludes,

> *Given the above, there is no perceived value to amending the language of Paragraph (a)(1)(ii)(B) to include atmospheric aboveground storage tanks if those storage tanks meet the provisions of NFPA 30.*

CHAPTER 4

OIL AND GAS-WELL DRILLING

I n Chapter 2 we noted that the Deepwater Horizon (DWH) catastrophe in the year 2010 led to the formation of the offshore regulatory agency BSEE (the Bureau of Safety and Environmental Enforcement). BSEE then issued its SEMS (Safety and Environmental Management Systems) rule, which, as we have seen, has strong similarities to the OSHA PSM standard. However, BSEE faces a challenge that OSHA has not had to consider — Deepwater Horizon was a drilling rig, not a production platform.

The offshore oil and gas business consists of two activities that are fundamentally different from one another: drilling and production. When a company is drilling a well, conditions are inherently dynamic — as the drilling progresses so operating conditions will change — sometimes quite quickly. Moreover, there is always uncertainty. In spite of the seismic surveys, no one can be completely sure what the formation looks like or what it contains.

Once the well has been drilled, and assuming that it can produce sufficient oil or gas, a production platform replaces the drilling rig.

Activities on the platform are analogous to an onshore refinery. Oil, gas and produced water come from the well. The three streams are separated and treated on the platform. The oil and gas are then pumped to an onshore location for further treatment. In other words, an offshore production platform is, in effect, a small, front-end refinery. The SEMS regulation is oriented toward production activities, not drilling operations. Which is why the SEMS and OSHA standards are so similar to one another. However, SEMS is less useful when it comes to drilling operations. For example, the SEMS/OSHA Management of Change (MOC) procedures are fully applicable to production activities, but they do not fit so well into a drilling environment.

THE REGULATION

Since OSHA has not assigned this proposed update to any particular category, it is shown here are being to do with the Application section.

(a)   *Application*
(b)   Definitions
(c)   Employee Participation
(d)   Process Safety Information
(e)   Process Hazards Analysis
(f)   Operating Procedures
(g)   Training
(h)   Contractors
(i)   Prestartup Safety Review
(j)   Mechanical Integrity
(k)   Hot Work
(l)   Management of Change
(m)   Incident Investigation

(n)   Emergency Planning and Response

(o)   Compliance Audits

(p)   Trade Secrets

Oil and gas-well drilling is not included in the current regulation. However, OSHA has made attempts to regulate onshore drilling operations. (The Chemical Safety Board has been involved in the follow up to some offshore drilling incidents, including Deepwater Horizon, but there has been dispute as to whether it has the authority to do so.)

PROPOSED UPDATE

Expanding the scope to include oil-and gas-well drilling and servicing.

DISCUSSION

In Chapter 2 we noted that recommendations from the Chemical Safety Board (CSB) are one of the reasons that OSHA is updating the process safety standard. One incident that the CSB investigated was the fatal fire at the Pryor Trust well (Chemical Safety Board, 2019). Five workers were killed. The following is from the Executive Summary to do with this incident.

On January 22, 2018, a blowout and rig fire occurred at Pryor Trust 0718 gas well number 1H-9, located in Pittsburg County, Oklahoma. The fire killed five workers, who were inside the driller's cabin on the rig floor. They died from thermal burn injuries and smoke and soot inhalation. The blowout occurred about three-and-a-half hours after removing drill pipe ("tripping") out of the well.

The cause of the blowout and rig fire was the failure of both the primary barrier—hydrostatic pressure produced by drilling mud—and the secondary barrier—human detection of influx and activation of the blowout preventer—which were intended to be in place to prevent a blowout.

The CSB issued nineteen recommendations, one of which was directed to OSHA. It reads,

Implement one of the three following options regarding regulatory changes:

a. *OPTION 1: Apply the Process Safety Management (PSM) standard (29 CFR 1910.119) to the drilling of oil and gas wells; or*

b. *OPTION 2: Apply the Process Safety Management (PSM) standard (29 CFR 1910.119) to the drilling of oil and gas wells as in OPTION 1, and make the necessary modifications to customize it to oil and gas drilling operations; or*

c. *OPTION 3: Develop a new standard with a safety management system framework similar to PSM that applies only to the drilling of onshore oil and gas wells that includes but is not limited to the following:*

1. *Detailed written operating procedures with specified steps and equipment alignment for all operations;*

2. *Written procedures for the management of changes (except replacements in kind) in procedures, the well plan, and equipment;*

3. *A risk assessment of hazards associated with the drilling plan;*

4. *A requirement to follow Recognized and Generally Accepted Good Engineering Practices (RAGAGEP);*

5. *Development of a Well Construction Interface Document between the operator and the drilling contractor prior to the commencement of drilling activities which at a minimum includes a bridging document and well plan specifying barriers and how to manage them;*

6. *The performance and documentation of flow checks using acceptable methods at defined points during the operation for a specified duration; and*

7. *A requirement for employee participation, similar to the Employee Participation requirement in the OSHA PSM standard.*

OPTION 1

As already discussed, there are fundamental differences between oil and gas drilling and the chemical process industries. Drilling occurs in a dynamic environment, whereas most chemical plants and refineries are steady state operations. Therefore, the language of the current process safety standard does not cover all aspects of drilling operations. Hence Option 1 is not really feasible.

OPTIONS 2 AND 3

Options 2 and 3 are similar to one another. The CSB is saying that the oil and gas–drilling industry is so different from chemical plants and oil refineries that they need their own standard.

A POSSIBLE OPTION 4

There is a fourth option that OSHA and BSEE may wish to con-sider. It may make sense to create a standard for both offshore and onshore drilling. That standard, which would have its roots in both PSM and SEMS, could be shared by OSHA and BSEE.

CHAPTER 5

OIL AND GAS PRODUCTION FACILITIES

The previous chapter discussed the difficulties to do with applying OSHA's process safety standard to oil and gas-drilling. The use of the standard for oil and gas production facilities is more straightforward because oil and gas production is a mostly steady-state activity.

THE REGULATION

Since OSHA has not assigned this proposed update to any particular category, it is shown here are being to do with the Application section.

(a)   *Application*
(b)   Definitions
(c)   Employee Participation
(d)   Process Safety Information
(e)   Process Hazards Analysis
(f)   Operating Procedures
(g)   Training

(h)   Contractors

(i)   Prestartup Safety Review

(j)   Mechanical Integrity

(k)   Hot Work

(l)   Management of Change

(m)   Incident Investigation

(n)   Emergency Planning and Response

(o)   Compliance Audits

(p)   Trade Secrets

The current regulation does not contain any specific provisions to do with oil and gas production. There are, however, various guidance documents (some of which are now archived) to do with this topic. Also, the SEMS (Safety and Environmental Management Systems) regulation from BSEE is very similar to OSHA's process safety standard.

PROPOSED UPDATE

The proposed update reads as follows,

> *Resuming enforcement for oil and gas production facilities*

SURFACE / SUB-SURFACE

A theme that runs through so many of the proposed updates to the standard is to do with boundaries — where does a covered process end? We see the same difficulty with regard to oil and gas-production. Does the standard apply just to activities conducted above ground, *i.e.*, downstream of the Christmas Tree? Or does it also apply to sub-surface operations?

ARCHIVE DOCUMENT

OSHA published an archive document (Occupational Safety and Health Administration, 1999) to do with oil and gas production facilities. Because it is archived, the document "may no longer represent OSHA policy." Nevertheless, it is useful to see what the agency had to say in the year 1999. The following is from that document.

> **Question**
>
> *Does the PSM standard (29 CFR 1910.119) apply to oil and gas production facilities, including oil, gas, and water separation facilities operating in conjunction with the producing well?*
>
> **Response**
>
> *. . . OSHA has stated in previous interpretation letters that production facilities, including related oil, gas, and water separation facilities, are excluded from PSM coverage under the oil and gas well drilling and servicing exemption, 29 C.F.R. §1910.119(a)(2) (ii). Several factors, however, demonstrate that the conclusions reached in these letters are erroneous. As a result, these letters are hereby rescinded.*
>
> *The letters in question fail to take into account the distinction between wells in production and those undergoing initial drilling or in a servicing status. Production, as recognized by the petroleum industry, is a phase of well operations that deals with bringing well fluids to the surface, separating them,*

and then storing, gauging and otherwise preparing the product for the pipeline. This production phase occurs after a well has been drilled, completed, and placed into operation, or after it has been returned to operation following workover or servicing. A completed well includes a "Christmas tree" (control valves, pressure gauges and choke assemblies to control the flow of oil and gas) which is attached at the top of the well where pressure is expected. It is at this point, the top of the well, where the covered PSM process begins. The distance between separation equipment and the well is not a factor when determining PSM applicability for production facilities.

CHAPTER 6

REACTIVE CHEMICAL HAZARDS

And, as we saw in Chapter 1, the preamble to the standard includes the following statement,

> *This section contains requirements for preventing or minimizing the consequences of catastrophic releases of toxic, reactive, flammable, or explosive chemicals. These releases may result in toxic, fire or explosion hazards.*

The topic of reactive chemical hazards is not called out explicitly in the standard, although its existence can certainly be implied.

The Chemical Safety Board has a particular concern with regard to 'reactive chemical hazards', as can be seen from the following quotation.

> *Reactive hazards are the dangers associated with uncontrolled chemical reactions in industrial processes. These uncontrolled reactions – such as thermal runaways and chemical decompositions –*

have been responsible for numerous fires, explosions, and toxic gas releases. From 1980 to 2001, 167 serious reactive accidents caused 108 fatalities in the U.S., according to the CSB's reactive hazards investigation.

Some of the incidents that the CSB cites are listed in Chapter 2.

THE REGULATION

The following are the elements of the current process safety standard. The topic of this chapter is to do with the first of these: *Application.*

(a) ***Application***
(b) Definitions
(c) Employee Participation
(d) Process Safety Information
(e) Process Hazards Analysis
(f) Operating Procedures
(g) Training
(h) Contractors
(i) Prestartup Safety Review
(j) Mechanical Integrity
(k) Hot Work
(l) Management of Change
(m) Incident Investigation
(n) Emergency Planning and Response
(o) Compliance Audits
(p) Trade Secrets

The current regulation does not contain any specific provisions to do with oil and gas production. There are, however, various guidance documents (some of which are now archived) that have been issued to do with this topic.

PROPOSED UPDATE

The proposed update for this topic is,

> *Expanding PSM coverage and requirements for*
> *reactive chemical*
> *hazards.*

OSHA does not provide any more information. Presumably, they plan to increase the number of chemicals included in Appendix A. What the other "requirements" may be are not identified. However, OSHA does address reactive chemical hazards in many of its standards, including the booklet, *Chemical Reactivity Hazards* and *Chemical Hazards and Toxic Substances*.

CHAPTER 7

HIGHLY HAZARDOUS CHEMICALS

The term 'highly hazardous chemicals' is included in the title of the standard. OSHA's definition for the term is,

> *Highly hazardous chemical means a substance possessing toxic, reactive, flammable, or explosive properties and specified by paragraph (a)(1) of this section.*

Appendix A of the standard (see Chapter 1) lists those chemicals that OSHA has identified as being highly hazardous.

THE REGULATION

The following are the elements of the current process safety standard. The topic of this chapter is to do with the second of these: *Definitions*.

(a)   Application
*(b)   **Definitions***

(c)   Employee Participation
(d)   Process Safety Information
(e)   Process Hazards Analysis
(f)   Operating Procedures
(g)   Training
(h)   Contractors
(i)   Prestartup Safety Review
(j)   Mechanical Integrity
(k)   Hot Work
(l)   Management of Change
(m) Incident Investigation
(n)   Emergency Planning and Response
(o)   Compliance Audits
(p)   Trade Secrets

PROPOSED UPDATE

Updating and expanding the list of highly hazardous chemicals in Appendix A.

OSHA has not provided a list of the chemicals that it proposes to add to Appendix A.

DISCUSSION

In the previous chapter to do with reactive chemical hazards we noted that the update to the standard has been strongly influenced by Chemical Safety Board recommendations. The CSB has also influenced the selection of highly hazardous chemicals. These are chemicals that are not included in the current Appendix A but that have been involved in serious incidents.

CHAPTER 8

EXPLOSIVES AND BLASTING AGENTS

Explosives are already covered by the process safety standard, as can be seen by the following two statements.

> *This section contains requirements for preventing or minimizing the consequences of catastrophic releases of toxic, reactive, flammable, or explosive chemicals. These releases may result in toxic, fire or explosion hazards.*

> *Highly hazardous chemical means a substance possessing toxic, reactive, flammable, or explosive properties and specified by paragraph (a)(1) of this section.*

However, OSHA is recommending that updates to do with explosive and blasting agents be handled at another of its standards: 29 CFR 1910.109.

THE REGULATION

The following are the elements of the current process safety standard. The topic of this chapter is to do with the second of these: *Definitions*.

(a)   Application
(b)   **Definitions**
(c)   Employee Participation
(d)   Process Safety Information
(e)   Process Hazards Analysis
(f)   Operating Procedures
(g)   Training
(h)   Contractors
(i)   Prestartup Safety Review
(j)   Mechanical Integrity
(k)   Hot Work
(l)   Management of Change
(m)   Incident Investigation
(n)   Emergency Planning and Response
(o)   Compliance Audits
(p)   Trade Secrets

The agency's definition of an explosive is,

> . . . *any chemical compound, mixture, or device, the primary or common purpose of which is to function by explosion...unless such compound, mixture, or device is otherwise specifically classified by the U.S Department of Transportation.*
>
> *Explosives also "include all material which is classified*

as Class A, Class B, and Class C explosives by the U.S Department of Transportation."

OSHA defines pyrotechnics as,

> *. . . any combustible or explosive compositions or manufactured articles designed and prepared for the purpose of producing audible or visible effects which are commonly referred to as fireworks.*

PROPOSED UPDATE

Amending paragraph (k) of the Explosives and Blasting Agents Standard (§ 1910.109) to extend PSM requirements to cover dismantling and disposal of explosives and pyrotechnics

DISCUSSION

29 CFR 1910.109 — Explosives and blasting agents — contains rules to do with storage of explosives, including information on safe distances, barricades, signage and the use of vehicles.

OSHA has published guidance documents *Process Safety Management for Explosives and Pyrotechnics Manufacturing* (Occupational Safety and Health Administration, 2017), and *Process Safety Management for Storage Facilities* (Occupational Safety and Health Administration, 2017).

The agency notes that, while all elements of the PSM standard apply to covered pyrotechnics manufacturing and storage facilities, the following elements are most relevant to hazards associated at these facilities:

- Employee Participation,
- Process Safety Information,
- Process Hazard Analysis,
- Operating Procedures,
- Training,
- Mechanical Integrity, and
- Emergency Planning and Response.

CHAPTER 9

CLARIFYING THE SCOPE OF THE RETAIL FACILITIES EXEMPTION

Many owners and managers of retail facilities store and sell hazardous chemicals. The owners and managers of these stores may be surprised to learn that their operations are part of the process industries, and that therefore they need to meet the requirements of a process safety management standard. As discussed in Chapter 1, the 10,000 lb. threshold for flammable and toxic materials is easily exceeded. Retail facilities can inadvertently cross that threshold all to easily.

Originally the PSM standard did not apply to retail facilities that sold hazardous chemicals to end users in small quantities. Therefore, gas stations, hardware stores, and other retailers were exempt from meeting the rule's requirements. However, OSHA was forced to issue letters of interpretation explaining how the exemption was to be understood. These statements culminated in the so-called '50 percent test', which stipulated that any establishment that obtained more than half of its income from direct sales to end users was exempt from PSM requirements.

This test had its own problems. Therefore, OSHA decided to follow the original intent of the NAICS (The North American Industry Classification System) Manual which states that, "Only facilities, or the portions of facilities, engaged in retail trade as defined by the current and any future updates to sections 44 and 45 of the NAICS Manual may be afforded the retail exemption." (The NAICS is the standard used by Federal statistical agencies in classifying business establishments for the purpose of collecting, analyzing, and publishing statistical data related to the U.S. business economy.

THE REGULATION

The following are the elements of the current process safety standard. The topic of this chapter is to do with the second of these: *Definitions,* which states: *This section does not apply to: (i) Retail facilities.*

(a)	Application
(b)	*Definitions*
(c)	Employee Participation
(d)	Process Safety Information
(e)	Process Hazards Analysis
(f)	Operating Procedures
(g)	Training
(h)	Contractors
(i)	Prestartup Safety Review
(j)	Mechanical Integrity
(k)	Hot Work
(l)	Management of Change
(m)	Incident Investigation
(n)	Emergency Planning and Response

(o)   Compliance Audits

(p)   Trade Secrets

PROPOSED UPDATE

The proposed update reads,

> *Clarifying the scope of the retail facilities exemption*

DISCUSSION

In 2018 OSHA issued a memorandum *Process Safety Management Retail Exemption Policy* (Occupational Safety and Health Administration, 2018). It contains the following statement,

> *OSHA's process safety management (PSM) standard, which contains requirements for preventing or minimizing toxic, fire, and explosion hazards associated with catastrophic releases of toxic, reactive, flammable, or explosive chemicals, does not apply to "retail facilities." The PSM standard does not define the term "retail," and on September 23, 2016, the United States Court of Appeals for the District of Columbia Circuit invalidated a memo stating OSHA's interpretation of that term.*

> *With respect to the exclusion of retail facilities ... OSHA believed that such facilities did not present the same degree of hazard to employees as other workplaces covered by the proposal. Therefore, OSHA should not require a comprehensive process*

safety management system in addition to other applicable OSHA standards addressing flammable and combustible liquids, compressed gases, hazard communication, etc., for retail facilities...

Certain highly hazardous chemicals may be present in [retail] ... operations. However, OSHA believes that chemicals in retail facilities are in small volume packages, containers and allotments, making a large release unlikely. OSHA received few comments disagreeing with the exemption of retail facilities (e.g., gasoline stations). OSHA has retained the exemption in the final rule.

In other words, there appears to be confusion as to what constitutes a retail operation, and whether such operations are to be covered by the process safety standard. Clarification is called for.

When first enacted in 1992, PSM standards did not apply to retail facilities that sold hazardous chemicals to end users in small quantities. As such, gas stations, hardware stores, and other purveyors of potentially harmful chemicals were exempt. But because the guidelines were considered ambiguous, OSHA was forced to issue numerous letters of interpretation explaining how the exemption was to be understood. These statements culminated in the so-called "50 percent test," which stipulated that any establishment that obtained more than half of its income from direct sales to end users was exempt from PSM requirements.

In time, however, OSHA decided that the new

standard was incompatible with the original intent of the chemical compliance exemption. In particular, the agency noted that many non-retail facilities had incorrectly utilized the exemption. Another troubling repercussion of the 50 percent test was that it allowed establishments to distribute huge quantities of chemicals directly to end users, even if they were commercial customers.

. . . new changes will nullify all of the earlier chemical compliance statements regarding the 50 percent test and the retail exemption.

(HSI, 2022)

In an archived document (*i.e.,* one that has been withdrawn), OSHA says the following,

OSHA considers an establishment to be qualified under the PSM "retail facilities" exemption, if that establishment receives more than half of its income from the direct sales of the PSM-covered highly hazardous chemical (HHC) to end users. The income referenced above applies to the income obtained from the sales of PSM-covered HHCs, and not the total sales of the establishment.

CHAPTER 10

DEFINING THE LIMITS OF A PSM-COVERED PROCESS

As we have seen in previous chapters, defining the limits of PSM-covered process is something that many of the proposed updates seek to address in one way or another. The decision as to where process safety boundaries should be located is both important and difficult.

THE REGULATION

(a)   Application

(b)   ***Definitions***

(c)   Employee Participation

(d)   Process Safety Information

(e)   Process Hazards Analysis

(f)   Operating Procedures

(g)   Training

(h)   Contractors

(i)   Prestartup Safety Review

(j)   Mechanical Integrity

(k) Hot Work
(l) Management of Change
(m) Incident Investigation
(n) Emergency Planning and Response
(o) Compliance Audits
(p) Trade Secrets

This topic can be placed into the Definitions section, which contains the following paragraphs.

i. *A process which involves a chemical at or above the specified threshold quantities listed in appendix A to this section;*

ii. *A process which involves a flammable liquid or gas (as defined in 1910.1200(c) of this part) on site in one location, in a quantity of 10,000 pounds (4535.9 kg) or more except for:*

(a) *Hydrocarbon fuels used solely for workplace consumption as a fuel (e.g., propane used for comfort heating, gasoline for vehicle refueling), if such fuels are not a part of a process containing another highly hazardous chemical covered by this standard;*

(b) *Flammable liquids stored in atmospheric tanks or transferred which are kept below their normal boiling point without benefit of chilling or refrigeration.*

PROPOSED UPDATE

Defining the limits of a PSM-covered process.

DISCUSSION

Concerns to do with what constitutes a PSM-covered process occur in many of the proposed updates. For example, both the Atmospheric Tanks and Retail Facilities updates are to do with "what's in" and "what's out". In fact, deciding on the limits of a process that is covered by the standard is a thread that runs through many of the proposed updates.

GUIDANCE

OSHA has offered the following guidance on this topic.

> *"Process" means any activity involving a highly hazardous chemical including any use, storage, manufacturing, handling, or the on-site movement of such chemicals, or combination of these activities. For purposes of this definition, any group of vessels which are interconnected and* **separate vessels which are located such that a highly hazardous chemical could be involved in a potential release** *shall be considered a single process.*
>
> *< my emphasis >*

In another letter, OSHA says,

> *For co-located equipment to be considered a PSM-covered process, there must be a reasonable probability that an event such as an explosion would affect unconnected vessels which contain quantities of the chemical that when added together would exceed the threshold quantity and provide a potential*

for a catastrophic release. In general, unconnected vessels must be evaluated by the employer to determine if they would interact during an incident, and if such a reasonable condition exists these vessels would be included in the process.

In some cases, physical barriers can effectively separate unconnected vessels to the extent that the potential for a chemical release is adequately controlled. For example, where a dike is used around a liquid storage vessel to fully contain released material and prevent it from interacting with another vessel outside the dike, and neither vessel by itself contains the threshold quantity, then this physical barrier would be considered acceptable in making the two vessels remote from each other. In addition, OSHA has previously stated that employers can use passive controls depending on the hazard to achieve adequate separation of vessels. Conversely, OSHA has also stated that other engineering controls (e.g., sprinkler systems, automatic-closing fire doors, etc.), and administrative controls (e.g., operating procedures, mechanical integrity procedures, and training) used to prevent and mitigate a catastrophic release of a covered HHC may not be used to determine the extent of a process as defined in paragraph 1910.119(b). As you have mentioned the presence of sprinklers in your facility, please note that building sprinklers are not a consideration for determining the boundaries of the covered process because they are an active engineering control.

Further guidance to do with process limits is provided by OSHA in a 1998 letter entitled Akzo-Nobel Chemicals – Limits of a Process (Occupational Safety and Health Administration, 1997).

THE REFINERY OPERATOR'S STORY

During a HAZOP on a large oil refinery, one of the senior operators said, "Just by opening and closing valves I can put gasoline into the refinery manager's coffee cup". There was little doubt that he was telling the truth. The point of this story is to demonstrate that almost any process can be connected to any other process.

On a more serious note, your author was involved in the follow-up to an incident in which a process line was connected to a utility system. It was only when the unit had a fire that they discovered that there were large quantities of gasoline in the fire water header. They came close to fighting a fire with 'literal fire water'.

INTERFACE HAZARDS ANALYSES

When it comes to process operations, we have seen that, in the limit, everything is connected to everything else.

This connectivity creates difficulties with process hazards analyses (PHAs). For example, the safety of a process facility that is supplied with electricity from an outside source depends on the reliability of that supplier. Yet the managers at the process facility cannot conduct hazards analyses at the utility. In this situation, what a company can do is carry out an Interface Hazards Analysis (IHA). The hazards analysis team can consider issues such as 'no flow', 'reverse flow' and 'failure to communicate' with regard to process flow, instrument signals and human communication (both at the working level, and between the respective managers).

Such a system can be viewed as being a collection of black boxes, where each black box represents an operating unit, each of which has been thoroughly analyzed individually. These black boxes are like nodes in a PHA.

CHAPTER 11

RAGAGEP: DEFINITION AND UPDATES

Note:
The proposed change 'Updates to RAGAGEP' (Chapter 14) is combined with the contents of this chapter. The contents of this chapter also share material with Chapter 19 (Mechanical Integrity of Critical Equipment) and Chapter 20 (Equipment Deficiencies).

OSHA, the EPA (Environmental Protection Agency), and the American Chemistry Council (ACC) promote the concept of RAGAGEP (Recognized and Generally Accepted Good Engineering Practice). The ACC includes RAGAGEP in its Responsible Care Process Safety Code. The EPA information is available at their web site (Environmental Protection Agency, 2022).

THE REGULATION

The following are the elements of the current process safety standard. RAGAGEP is to do with three of these: Definitions, Process Safety Information and Mechanical Integrity.

(a)   Application
(b)   Definitions
(c)   Employee Participation
(d)   Process Safety Information
(e)   Process Hazards Analysis
(f)   Operating Procedures
(g)   Training
(h)   Contractors
(i)   Prestartup Safety Review
(j)   Mechanical Integrity
(k)   Hot Work
(l)   Management of Change
(m) Incident Investigation
(n)   Emergency Planning and Response
(o)   Compliance Audits
(p)   Trade Secrets

The standard refers the topic of RAGAGEP in three of its provisions:

(d)(3)(ii): Employers must document that all equipment in PSM–covered processes complies with RAGAGEP;

(j)(4)(ii): Inspections and tests are performed on process equipment subject to the standard's mechanical integrity requirements in accordance with RAGAGEP; and

(j)(4)(iii): Inspection and test frequency follows manufacturer's recommendations and good engineering practice, and more frequently if indicated by operating experience.

In addition, paragraph (d)(3)(iii) addresses situations where the design codes, standards, or practices used in the design and construction of existing equipment are no longer in general use. In such cases, the employer must determine and document that the equipment is designed, maintained, inspected, tested, and operating in a safe manner.

PROPOSED UPDATES

The two proposed updates to do with RAGAGEP are shown below.

> *Amending paragraph (d) to require evaluation of updates to applicable recognized and generally accepted as good engineering practices (RAGAGEP).*
>> Paragraph (d) is to do with Process Safety Information.

> *Amending paragraph (b) to include a definition of RAGAGEP;*
>> Paragraph (b) is to do with definitions.

OSHA GUIDANCE

In May 2016, OSHA issued the guidance document *RAGAGEP in Process Safety Management Enforcement*. This is a lengthy document. The *Enforcement Considerations* section alone contains 16 paragraphs. The following are the first four points from that section. (Occupational Safety and Health Administration, 2016).

1. There may be multiple RAGAGEP(s) that apply to a specific process. For example, American Petroleum Institute (API), RP 520 Sizing, Selection, and Installation of Pressure-Relieving Devices in Refineries Part II - Installation, and International Standards Organization, Standard No. 4126-9, Application and installation of safety devices, are both RAGAGEP for relief valve installation and contain similar but not identical requirements. Both documents are protective and either is acceptable to OSHA.

2. Employers do not need to consider or comply with a RAGAGEP provision that is not applicable to their specific worksite conditions, situations, or applications.

3. Some employers apply RAGAGEP outside of their intended area of application, such as using ammonia refrigeration pressure vessel inspection recommended practices in a chemical plant or refinery process. Use of inapplicable RAGAGEP can result in poor hazard control and can be grounds for citations.

4. There may be cases where the selected RAGAGEP does not control all of the hazards in an employer's covered process. As discussed above, the employer is expected to adopt other RAGAGEP (potentially including internal standards, guidance, or procedures) to address remaining process hazards. Whether internal standards constitute RAGAGEP should be reviewed on a case-by-case basis.

Further guidance to do with RAGAGEP is provided in the ebook *Asset Integrity* (Sutton, Sutton Technical Books, 2017).

DISCUSSION

RAGAGEP basically has two components: Compliance and Judgment. The compliance element is fairly straightforward. RAGAGEP requires full compliance with all pertinent codes, standards and professional guidance documents. Given the enormous number of these items, keeping up with them all is no mean task.

RAGAGEP also involves judgment in those situations where a code or standard does not exist, or where the standard is ambiguous or open to interpretation. Engineers and designers are called on to determine if a particular design decision 'makes sense' for the context in which it is being used. They have to be particularly careful that an engineering decision that improves production or productivity does not degrade safety or environmental performance.

Like so many other process safety elements, there is a danger of falling into the trap of self-referential reasoning, as discussed in Chapter 27. What constitutes 'Recognized' and 'Generally Accepted'? There are bound to be differences of opinion between experts. It is all too easy to fall into circular logic of the following form.

- Who defines "recognized and good engineering practice"?
- Engineering experts.
- What makes them engineering experts?
- They establish "recognized and good engineering practice".

One important benefit of using RAGAGEP is that process safety limits (see Figure 20.1) can be defined more tightly.

RAGAGEP DEVELOPMENT

The development of RAGAGEPs for a particular company or facility generally includes the following steps:

1. Identify the relevant federal and state regulations,
2. Identify local codes and standards (such as building and fire codes),
3. Identify the pertinent industry consensus standards,
4. Review all of the above with legal, safety and environmental staff,
5. Incorporate proprietary experience and standards, and
6. Finalize with engineering judgment.

CHAPTER 12

DEFINITION OF CRITICAL EQUIPMENT

O n a large process facility many items of equipment and instrumentation will not be performing optimally. Indeed, there could be hundreds of items that require attention in one way of another. However, only a few of these items can be considered to ber safety critical, *i.e.*, were they to fail a serious incident could ensue. Some means of determining criticality is needed.

One way of identifying critical items is to conduct a systems analysis using a technique such as fault tree analysis. This will help differentiate the "important few" from the "unimportant many" (Sutton, Safety Moment #94: Fault Tree Analysis, 2020), (Sutton, Frequency Analysis, 2018).

THE REGULATION

The following are the elements of the current process safety standard. The current regulation does not include any reference to the concept of Critical Equipment. OSHA has placed the proposed update into paragraph (b) — Definitions.

- (a) Application
- (b) *Definitions*
- (c) Employee Participation
- (d) Process Safety Information
- (e) Process Hazards Analysis
- (f) Operating Procedures
- (g) Training
- (h) Contractors
- (i) Prestartup Safety Review
- (j) Mechanical Integrity
- (k) Hot Work
- (l) Management of Change
- (m) Incident Investigation
- (n) Emergency Planning and Response
- (o) Compliance Audits
- (p) Trade Secrets

PROPOSED UPDATE

In its Stakeholder agenda (Occupational Safety and Health Administration, 2022) OSHA states that it is,

> *Amending paragraph (b) to include a definition of critical equipment*

(Paragraph (b) of the standard is to do with definitions.)

DISCUSSION

Broadly speaking, there are two types of critical items: critical equipment, and safety critical instrumentation.

CRITICAL EQUIPMENT

The Center for Chemical Process Safety (CCPS) offers the following definition for the term "critical equipment".

> *Equipment, instrumentation, controls, or systems whose malfunction or failure would likely result in a catastrophic release of highly hazardous chemicals, or whose proper operation is required to mitigate the consequences of such release. (Examples are most safety systems, such as area LEL monitors, fire protection systems such as deluge or underground systems, and key operational equipment usually handling high pressures or large volumes.)*

The offshore community uses the term 'Safety Critical Systems' when developing Safety Cases. They recognize that serious incidents generally require that failure of an overall system, including its layers of protection. Safety Critical Systems include all components: hardware, software and human actions. The failure of such a system can result in death or serious injury, or severe equipment and operating loss.

Some equipment, such as pressure safety relief valves, are obviously critical. However, it is not always possible to define what makes an equipment item critical. For example, the operating technicians may rely on an apparently insignificant pressure gauge to check that the pressure in a vessel is not too high. To an outsider this gauge may not seem to be critical at all. Yet, it is actually critical to safe operation.

A second difficulty to do with defining the term "critical equipment" is that most such items are do not operate in isolation. If a level instrument, for example, is critically important to safety, then it is likely that it will already be provided with a backup, or that the control system will use some other method of moving the system into a safe state. Therefore, the criticality applies to the whole system, not to just one instrument. (One way of identifying critical equipment would be to identify single-point failures that could lead to catastrophic consequences.)

SAFETY EQUIPMENT AND INSTRUMENTATION

Some safety equipment and instrumentation will always be safety critical. For example, a pressure vessel may be protected against an over-pressure situation as follows.

- *Control instrumentation*
  The normal pressure control instrumentation will respond before the situation becomes critical.
- *Operator Response*
  The operating technicians note that the pressure is rising, so they take action.
- *Operating Alarms*
  If action is not taken by the normal control instrumentation or the operators, a high pressure alarm will sound. This will bring the situation to the attention of the operators, who will be required to take action in order to remove the alarm signal.
- *Safety Critical Instrumentation*
  If the pressure in the vessel is still rising, the safety

critical system will take control. It will take the actions necessary to bring the situation to a safe state.

For safety instrumented systems, risk reduction is measured using values for Safety Integrity Level (SIL).

- **Pressure Safety Relief Valve**

  Finally, if all else has failed, a pressure safety relief valve will open, thus allowing the contents of the vessel to go to a safe location.

CHAPTER 13

EMPLOYEE PARTICIPATION AND STOP WORK AUTHORITY

Employee participation lies at the heart of any successful process safety management program. Indeed, participation can be considered the most important of the process safety elements. If all workers (not just direct employees) are fully involved in the safety management process, and if they feel empowered to make suggestions for improvement, then the facility is likely to have a good safety record. These employees are also much more likely to make suggestions for improvements, to participate in new initiatives and to "walk the extra mile" when called upon. Moreover, the effective involvement of the workforce provides a sanity check for new ideas, projects and analyses. Anything new or unusual should be reviewed by the employees; they will immediately identify any common sense problems because they are the ones who know the facility best.

THE REGULATION

(a) Application
(b) Definitions
(c) *Employee Participation*
(d) Process Safety Information
(e) Process Hazards Analysis
(f) Operating Procedures
(g) Training
(h) Contractors
(i) Prestartup Safety Review
(j) Mechanical Integrity
(k) Hot Work
(l) Management of Change
(m) Incident Investigation
(n) Emergency Planning and Response
(o) Compliance Audits
(p) Trade Secrets

Paragraph (c) of the OSHA standard reads as follows.

> *(1) Employers shall develop a written plan of action regarding the implementation of the employee participation required by this paragraph.*
>
> *(2) Employers shall consult with employees and their representatives on the conduct and development of process hazards analyses and on the development of the other elements of process safety management in this standard.*
>
> *(3) Employers shall provide to employees and their*

representatives access to process hazard analyses and to all other information required to be developed under this standard.

PROPOSED UPDATE

Expanding paragraph (c) to strengthen employee participation and include stop work authority.

GUIDANCE

The following guidance to do with employee participation is provided by OSHA.

Section 304 of the Clean Air Act Amendments states that employers are to consult with their employees and their representatives regarding the employers efforts in the development and implementation of the process safety management program elements and hazard assessments. Section 304 also requires employers to train and educate their employees and to inform affected employees of the findings from incident investigations required by the process safety management program. Many employers, under their safety and health programs, have already established means and methods to keep employees and their representatives informed about relevant safety and health issues and employers may be able to adapt these practices and procedures to meet their obligations under this standard. Employers who have not implemented an occupational safety and health program may wish to form a safety and health committee of employees and management representatives to help the employer meet the obligations specified by this standard. These committees can become a significant ally in helping the employer to implement and maintain an effective process safety management program for all employees.

DISCUSSION

OSHA has provided no detail as to what they meant by "strengthening the standard". The following material, which is taken from the book *Process Risk and Reliability Management* (Sutton, Process Risk and Reliability Management, 2014), provides some suggestions.

NOT JUST COMMUNICATION

It is important to note that this element is called Employee *Participation,* not Employee *Communication.* The intent is that employees fully engage in the spirit of the process safety program. Moreover, the word 'employee' should not be restricted to people who are actually on the company's payroll — it should include contract workers, consultants and visitors.

PROCESS HAZARDS ANALYSES (PHAS)

OSHA discusses the importance of employees participating in PHAs.

All employees should be encouraged to participate in the PHA meetings themselves. They should have a chance to contribute their knowledge, experience and ideas. By participating in PHAs they develop a process safety way of thinking — they will start to look at everything they do in terms of its risk impact, and they will become better at "thinking the unthinkable". For example, an operator working by himself at one o'clock in the morning may be about to open a valve on a line that connects two tanks. If, before doing so, he spends a few moments going through some of the PHA guidewords such as "Reverse Flow" or "Contamination" he may identify a possible accident situation and decide not to open the valve until he has talked over the proposed action with his supervisor or colleagues. When the operator acts in this manner both the participation and the PHA elements of the process safety program are working perfectly. Employee participation is not a stand-alone activity; instead it should be woven into the fabric of all the elements of a risk management program.

Additional examples of workforce involvement occur when a pipefitter learns that a new chemical is about to be used in the process. He may question whether the current gaskets are safe in the new service. Or an outside contractor may feel that he or she has not been given sufficient instructions as to what to do and where to go in an emergency, and makes that concern known to the host company.

PERSONAL REPUTATION

Although there are many benefits to do with participation, management has to recognize that, by asking employees to get involved in decision making they are also asking those employees to take more risk with regards to their career and reputation. It is much easier for an employee merely to follow orders — even if he or she knows that those orders are not sensible — than to take initiative.

Increased employee participation may run into roadblocks with unions and other organizations that represent those employees. Consequently, employees must feel that they are sufficiently rewarded for participating in management programs.

SAFETY COMMITTEES

Safety committees provide a formal channel through which management and the employees can communicate with regard to process safety issues and overall company culture. If the facility is non-union, it is essential that the employees' representative is selected by the employees, not appointed by management. But it is important to ensure that the committee is not made up of just a select few; the effective implementation of this element requires that *everyone* participate in the process safety program.

WRITTEN PLAN

Management and the employees should develop a written plan showing how they plan to implement Workforce Involvement. An example is shown in Table 13.1.

Table 13.1: Example Process Safety Written Plan

1.	The PSM program will involve all employees and contract workers, as appropriate to their job function and experience level.
2.	The program will involve the full participation of employee representatives — where such duly elected representatives exist.
3.	The term 'Employees' includes not only full-time workers, but also temporary, part-time and contract workers.

4.   Decisions as to which employees should be consulted re-garding specific PSM matters will take into account factors such as job functions, experience, and their degree of in-volvement with PSM and time with the company.

5.   Employees can participate in the PSM program by taking leadership of some of the elements of process safety. It is a good idea to involve employees with lower levels of experi-ence wherever possible in order to train them in the details of the process safety program.

STOP WORK AUTHORITY

OSHA is proposing to add stop-work authority to this element of the standard. As we have seen, the offshore community in the United States developed its own process safety standard (SEMS) following the Deepwater Horizon catastrophe. This standard (Bureau of Safe-ty and Environmental Enforcement, 2022), stressed the importance of providing workers with the authority to stop an activity if they felt that they were in imminent danger.

Based on a similar enhancement to the EPA's Risk Management Program (Environmental Protection Agency, 2022) OSHA may be considering a provision similar to the proposed RMP amendments. It would require employers to ensure that employees and their rep-resentatives have authority to refuse to perform a task that could reasonably result in a catastrophic release. This provision would also allow a qualified operator to partially or completely shut down an operation or process based on the potential for a catastrophic release.

Although the basic idea behind the proposed change makes sense, it is important to distinguish between *Stop Work* and *Stop the Pro-*

cess authority. As we have discussed in Chapters 4 and 5, drilling is a dynamic process. Conditions are constantly changing. The biggest risk is that of a well blowout. Many oil and gas wells are at very high pressure. Therefore, when drilling into them it is vital to ensure that the oil and gas does not flow up the drill string and on to the drill platform. If it does the oil and gas will probably ignite and the resulting fire could be catastrophic. The poster child for such a situation is the Deepwater Horizon catastrophe.

In a potential blowout situation, it is possible that the workers on the drilling rig will realize that the situation is badly awry. They can therefore make the (expensive) decision to manually shear the drill pipe with rams that are a part of the blowout preventer. The SEMS rule makes it clear that, if a worker takes that action, then he or she cannot be disciplined.

Based on the analyses provided in Chapter 4, situations such as these are not nearly as likely to occur on a steady-state operation such as a chemical plant or refinery. If a worker shuts down a continuously operating process suddenly, it is possible that the resulting situation could be more hazardous than had he or she done nothing. However, if the same worker observes an unsafe work activity, such as a crane swinging its load too close to process equipment, then it makes sense to stop that activity until the unsafe conditions are corrected.

CHAPTER 14

UPDATES TO RAGAGEP

The topic of RAGAGEP (Recognized and Generally Accepted Good Engineering Practice) has already been discussed in Chapter 11 — Definition of RAGAGEP. It is not clear as to why OSHA chose to mention this topic twice — the two proposed changes could have been merged. Therefore, discussion to do with this proposed change is provided in Chapter 11.

CHAPTER 15

UPDATING COLLECTED INFORMATION

Technical information is the bedrock of all other process safety activities. If the information is out of date serious mistakes can be made with hazards analyses, operating procedures, and many other critical activities. The term that OSHA uses — Process Safety Information — is too limiting. A better phrase would be Technical Information. It includes equipment data sheets, P&IDs, instrument loop diagrams, and safety data sheets. Not all of this information may be to do with safety explicitly, but it is all needed for safe and efficient operations.

THE REGULATION

(a) Application
(b) Definitions
(c) Employee Participation
(d) *Process Safety Information*
(e) Process Hazards Analysis
(f) Operating Procedures

(g)   Training

(h)   Contractors

(i)   Prestartup Safety Review

(j)   Mechanical Integrity

(k)   Hot Work

(l)   Management of Change

(m)   Incident Investigation

(n)   Emergency Planning and Response

(o)   Compliance Audits

(p)   Trade Secrets

The relevant section of the current regulation, from paragraph (d), reads,

> ... the employer shall complete a compilation of written process safety information before conducting any process hazard analysis required by the standard. The compilation of written process safety information is to enable the employer the employees involved in operating the process to identify and understand the hazards posed by those processes involving highly hazardous chemicals.

PROPOSED UPDATE

The proposed modification to paragraph (d) is,

> Amending paragraph (d) to require continuous updating of collected information.

It is not clear as to why OAHA introduced the word "collected" here. The update should apply to all types of technical information, regardless of its source.

DISCUSSION

This proposed update seems to be common sense. As process safety information becomes dated it will be less and less useful. Eventually there comes a point where an accident may occur because someone made a bad decision based on incorrect information. The fact that OSHA did not include a requirement for updating information in the original standard appears to be no more than a simple oversight.

UPDATE PROCESS

There needs to be a process that ensure that process safety information is updated on a timely basis. There also needs to be a process by which those who use that information know that what they have is current and up to date.

The proposed update does not say how often the process safety information needs to be updated. Conditions change on most facilities change on an almost a daily basis, and many of those changes require that the technical information base be updated. It is not feasible to update the information instantaneously, so a policy for collecting, checking and formally updating process safety information is needed.

MANAGEMENT OF CHANGE

The Management of Change element of the regulation, paragraph (I), contains the following paragraph,

If a change covered by this paragraph results in a

change in the operating procedures or practices required by paragraph (f) of this section, such procedures or practices shall be updated accordingly.

This paragraph strongly implies that technical information has to be kept up to date because such information is a necessary foundation of accurate operating procedures.

That paragraph also states,

In this process safety management standard, change includes all modifications to equipment, procedures, raw materials and processing conditions other than "replacement in kind". These changes need to be properly managed by identifying and reviewing them prior to implementation of the change. For example, the operating procedures contain the operating parameters (pressure limits, temperature ranges, flow rates, etc.) and the importance of operating within these limits.

An effective Management of Change program should automatically ensure that technical information is updated, as needed, following any significant change to the equipment or process conditions.

REPLACEMENT IN KIND

When it comes to updating information, the topic of "replacement in kind" needs to be considered. If a piece of equipment is replaced with a new item that has different performance specifications, that is fabricated from different materials, or requires different maintenance procedures, then the relevant information needs to be updated quickly. If, however, an item is being replaced with something that is functionally identical, then there is less urgency when it comes to updating the information base.

CHAPTER 16

FORMAL RESOLUTION OF PHA RECOMMENDATIONS

P rocess Hazards Analysis (PHA) is at the core of any process safety program. It is essential to identify the hazards that are present, and then to assess their consequences and the likelihood of their occurrence. Based on the analyses, corrective action can be recommended.

When the process safety standard was still quite new, PHAs were a priority activity. Your author recalls talking to a chemical plant manager not long after OSHA had introduced its process safety management regulation in the early 1990s. The manager said, "I know what PSM is, it's HAZOPs!" In fact, the HAZOP method is just one of seven hazards analysis techniques which are themselves just one of the fourteen elements of the standard. Nevertheless, the manager did have a point. Unless he and his team could identify the hazards at the facility, they could not take corrective action.

THE REGULATION

(a)   Application
(b)   Definitions
(c)   Employee Participation
(d)   Process Safety Information
(e)   *Process Hazards Analysis*
(f)   Operating Procedures
(g)   Training
(h)   Contractors
(i)   Prestartup Safety Review
(j)   Mechanical Integrity
(k)   Hot Work
(l)   Management of Change
(m)   Incident Investigation
(n)   Emergency Planning and Response
(o)   Compliance Audits
(p)   Trade Secrets

The relevant section from paragraph (e) of the regulation reads,

> *The employer shall establish a system to promptly address the team's findings and recommendations; assure that the recommendations are resolved in a timely manner and that the resolution is documented; document what actions are to be taken; complete actions as soon as possible; develop a written schedule of when these actions are to be completed; communicate the actions to operating, maintenance and other employees whose work assignments are in the process and who may be affected by the recommendations or actions.*

PROPOSED UPDATE

The proposed change to paragraph (e) is,

Amending § 1910.119(e) to require formal resolution of PHA team recommendations that are not utilized

DISCUSSION

The success and importance of PHAs has created a problem that OSHA wants to see corrected — PHAs can identify hazards and recommendations which are not necessarily acted on. There are three reasons for this lack of follow up.

First, the identified hazard is considered to be low risk, so management judges that action is not required, at least not for now. The second reason for lack of follow-up is that the risk may be considered to be below the threshold value of acceptable risk. The third reason is that the follow-up action "gets lost" due to an administrative oversight or changes in management.

The proposed update is to do with the first of these problems — deferred action when risk is considered to be low. However, in practice, it often turns out that the third concern — administrative oversight — is more critical.

RISK EVALUATION

It is not enough for a PHA team simply to identify hazards — it must also assign a risk to each of those hazards. This is usually done through the use of risk matrices (Sutton, Process Risk and Reliability Management, 2014).

Commonly, four levels of risk are assigned to an identified hazard.

A — (Red) Very High
This level of risk requires prompt action; money is no object, and the option of doing nothing is not an option. An 'A' risk is urgent. On an operating facility, management must implement Immediate Temporary Controls (ITC) while longer-term solutions are being investigated. If effective ITCs cannot be found, then the operation must be stopped. During the design phases of a project immediate corrective action must be taken in response to an 'A' finding, regardless of the impact on the schedule and budget.

B — (Orange) High
Risk must be reduced, but there is time to conduct more detailed analyses and investigations. Remediation is expected within say 90 days. If the resolution is expected to take longer than this, then an ITC must be put in place.

C — (Yellow) Moderate
The risk is significant. However, cost considerations can be factored into the final action taken, as can normal scheduling constraints such as the availability of spare parts or the timing of facility turnarounds. Resolution of the finding must occur within say 18 months. An ITC may or may not be required.

D — (Green) Low
Requires action but is of low importance. In spite of their low risk ranking, 'D' level risks must be resolved and recommendations implemented according to a schedule; they cannot be ignored. (Some companies do allow very low risk-ranked findings to be

ignored on the grounds that they are within the bounds of acceptable risk.)

It is unlikely that items in the red or orange categories will be overlooked. The proposed update will usually be to do with items that are in either the yellow or green zones.

In order to address the requirements of the proposed update, management and the hazards analysis team need to consider the following questions and concerns.

- Who determines the level of risk such that action can be deferred or avoided?
- What gives management the authority to ignore, defer or over-ride a PHA team's recommendation? After all, a key feature of a PHA is that the team is made up of representatives from many different department and disciplines who pool their expertise to come up with a risk assessment. What gives a manager the right to over-ride this combined expertise?
- How does management address the vexed topic of acceptable risk?

THE FOLLOW-UP PROBLEM

The proposed OSHA update does not address a difficult question faced by managers charged with the follow up to a PHA finding: selection of the best response.

For example, the PHA team may determine that a tank containing a flammable liquid could overflow, leading to a potential fire or a serious environmental violation. The team determines that the risk is unacceptably high (it is in the 'A' or 'B' categories).

Therefore, therefore "something should be done". The manager charged with follow up has various options. These include:

- Upgrade the level control instrumentation.
- Rewrite the operating procedures, and then train the operating technicians in the use of those procedures.
- Change the process so that the tank is no longer needed.
- Change the process so that a less hazardous chemical can be used.
- . . .

The concern here is not that management will fail to take action. The concern is that management will not take the *most effective* action. The proposed update to the regulation does not address this difficulty.

THE ONE-MINUTE ENGINEERING DEPARTMENT

Typically, hazards analysis teams have many members who have a technical background. By training and instinct, these people have a tendency to want to solve problems, yet they must understand that the purpose of the analysis is to identify hazards, not to come up with solutions; a hazards analysis team is not a one-minute engineering department. The purpose of a hazards analysis is to identify hazards, and to assess the risks to do with those hazards. It is not the purpose of a hazards analysis to come up with solutions. (In practice, the team will often identify potential responses, but such insights are incidental to the team's fundamental goal, which is to identify hazards.)

In practice, once the PHA meetings are concluded some members of the team will probably be assigned some of the follow-up tasks. Yet everyone must clearly understand that the role of team member and that of engineering support are different.

An important reason for not having the hazards analysis team develop recommendations is that the mental process for finding hazards is quite different from that for solving problems. When finding hazards, the team is looking for problems and hazards. When generating specific recommendations, the team is in a problem-solving mode. The two thought processes are fundamentally different.

If the leader can keep a grip on the tendency of the team to want to solve problems, he or she will find that the meeting proceeds quickly — which will please everyone, particularly upper management.

CHAPTER 17

SAFER TECHNOLOGY AND ALTERNATIVES ANALYSIS

Paragraph (e) of the current standard is to do with Process Hazards Analysis (PHA). A core feature of this paragraph is that the PHA team analyzes the system *as is,* and determines if that system does or does not pose an acceptable level of risk. Even when analyzing a system that is still in the design phase, the team evaluates the process and equipment that have been selected by the design engineers. Looked at this way, a process hazards analysis takes the form of an audit.

OSHA has now introduced a subtle, but profound, change. With the proposed update the agency is calling on companies to evaluate not just 'what is', but 'what might be'. This change may not be quite so profound for teams that are analyzing a new process that has yet been built. Nevertheless, there is a shift in focus of the PHA element of the standard.

THE REGULATION

    (a)   Application
    (b)   Definitions
    (c)   Employee Participation
    (d)   Process Safety Information
    (e)   ***Process Hazards Analysis***
    (f)   Operating Procedures
    (g)   Training
    (h)   Contractors
    (i)   Prestartup Safety Review
    (j)   Mechanical Integrity
    (k)   Hot Work
    (l)   Management of Change
    (m) Incident Investigation
    (n)   Emergency Planning and Response
    (o)   Compliance Audits
    (p)   Trade Secrets

The relevant sections of the current regulation are subparagraphs (1) and (3) of paragraph (e) — Process hazards analysis.

> *(1) The employer shall perform an initial process hazard analysis (hazard evaluation) on processes covered by this standard. The process hazard analysis shall be appropriate to the complexity of the process and shall identify, evaluate, and control the hazards involved in the process.*

> *(3) The process hazard analysis shall address: (i) The hazards of the process; (ii) The identification of any previous incident which had a likely potential for*

catastrophic consequences in the workplace; (iii) Engineering and administrative controls applicable to the hazards and their interrelationships such as appropriate application of detection methodologies to provide early warning of releases. (Acceptable detection methods might include process monitoring and control instrumentation with alarms, and detection hardware such as hydrocarbon sensors.); (iv) Consequences of failure of engineering and administrative controls; (v) Facility siting; (vi) Human factors; and (vii) A qualitative evaluation of a range of the possible safety and health effects of failure of controls on employees in the workplace.

PROPOSED UPDATE

OSHA has proposed the following update to the Process Hazards Analysis paragraph.

> *Expanding paragraph (e) by requiring safer technology and alternatives analysis*

DISCUSSION

This proposed update is short, open-ended, ambiguous, and seem-ingly innocuous. But it would drastically change the scope and in-tent of the PHA. The core purpose of a process hazards analysis is to identify hazards, assess the associated risk, and recommend that action be taken with regard to high-risk hazards. As we saw in the previous chapter, it is *not* the purpose of a PHA to come up with solutions. Yet that is what this proposed update is driving toward. By "expanding paragraph (e) to require safer technology and alternatives analysis" OSHA has introduced a subtle, but profound, change. The agency is now calling on companies to evaluate not just 'what is', but 'what might be'. This change may not be quite so profound for teams that are analyzing a new process that is in the design state and that has yet been built. Nevertheless, there is a shift in focus of the PHA element of the standard.

In the previous chapter to do with Formal Resolution of PHA Recommendations — we showed how an apparently simple hazard — overflow of a tank — generated a wide range of potential responses. These responses included changes to technology, changes to the process or materials of construction, and modifications to the operating procedures and training. It is extremely unlikely that the PHA team has expertise in all these areas. Yet, if the proposed change is adopted, the team is now called upon to develop and evaluate new systems. The proposed change to the standard creates a number of difficulties, including the following,

- The PHA team is not a design team. The participants in the PHA are not likely to have the skills and project background to develop, design and engineer "safer technology and alternatives". As discussed in the previous chapter, a process hazards analysis team is not a one-minute engineering department.
- The word "safer" leads, as so often with process safety, to potential circularity. The only way to know if a new design is safer is to conduct a PHA on that design. There is no way of being sure that it is safer before the analysis.
- The word "alternatives" potentially opens up the development of new approaches to the items in subparagraphs (v) and (vi): facility siting, and human factors. It is unlikely that

the PHA team members have the expertise to develop new designs that consider these topics.

CHAPTER 18

NATURAL DISASTERS AND EXTREME TEMPERATURES

Most of the proposed updates to the standard fall are to do with tuning existing elements of the regulation. OSHA is tuning existing standards, taking care of legalistic difficulties, and addressing Chemical Safety Board concerns. However, some of the proposed changes could be a door that opens on to new ways of interpreting and applying the regulation. This element — Natural Disasters and Extreme Temperatures — is an example of that type of change.

THE REGULATION

(a) Application
(b) Definitions
(c) Employee Participation
(d) Process Safety Information
(e) *Process Hazards Analysis*
(f) Operating Procedures
(g) Training

(h)   Contractors
(i)   Prestartup Safety Review
(j)   Mechanical Integrity
(k)   Hot Work
(l)   Management of Change
(m) Incident Investigation
(n)   Emergency Planning and Response
(o)   Compliance Audits
(p)   Trade Secrets

Paragraph (e) — Process Hazards Analysis — of the standard does not contain explicit provisions to do with natural disasters. However, most PHA teams do discuss predictable natural events, such as local flooding or extended periods of freezing temperatures. The key word here is 'predictable' — the team will not usually discuss the impact of new topics such as climate change or loss of biosphere diversity. PHA teams generally confine their discussions to events that are considered plausible within current norms.

PROPOSED UPDATE

Clarifying paragraph (e) to require consideration of natural disasters and extreme temperatures in their PSM programs, in response to E.O. 13990

(E.O. 13990 is described in Chapter 2.)

DISCUSSION

Most of the proposed updates to the standard are short and tend to be rather cryptic. They do not go into detail. This paragraph is particularly enigmatic. In particular, it does not distinguish between those events that have already occurred, and that can reasonably be expected to repeat, and the hard-to-define calamities that are looming, but that are far from certain to take place.

For example, with regard to natural disasters, a process facility on the Texas Gulf Coast has probably experienced severe flooding and hurricanes. Hence, the people working at these facilities know how to respond to such events. However, they do not have experience of responding to extended droughts and the associated loss of fresh water supplies — events that may occur as the climate changes. It would have been helpful had OSHA defined the phrase 'natural disasters' in greater detail.

The term 'Extreme Temperatures' is even more challenging. What is meant by the word extreme? Is OSHA expecting companies to consider the impact of global temperatures reaching the level where massive social disruptions can be anticipated? Or are they considering a narrower meaning confined to maintaining sufficiently low cooling water temperatures? They provide no guidance.

Another difficulty is that climate change is not a single event. It occurs on a continuum. We are already experiencing some 'extreme temperatures'. But temperatures will continue to climb. At what point do we draw the line?

CHAPTER 19

MECHANICAL INTEGRITY OF CRITICAL EQUIPMENT

OSHA wants to ensure that the mechanical integrity of critical equipment and instrumentation is properly managed. Yet, in Chapter 12 we saw that determining which items are critical is a difficult judgment to make.

THE REGULATION

The pertinent section of the regulation is paragraph (j) — Mechanical Integrity.

- (a) Application.
- (b) Definitions
- (c) Employee Participation
- (d) Process Safety Information
- (e) Process Hazards Analysis
- (f) Operating Procedures
- (g) Training
- (h) Contractors
- (i) Prestartup Safety Review

(j)   ***Mechanical Integrity***

(k)   Hot Work

(l)   Management of Change

(m)   Incident Investigation

(n)   Emergency Planning and Response

(o)   Compliance Audits

(p)   Trade Secrets

The following is from the relevant section of paragraph (j).

> *Paragraphs (j)(2) through (j)(6) of this section apply to the following process equipment:*
>
> (i)   *Pressure vessels and storage tanks;*
>
> (ii)   *Piping systems (including piping components such as valves);*
>
> (iii)   *Relief and vent systems and devices;*
>
> (iv)   *Emergency shutdown systems;*
>
> (v)   *Controls (including monitoring devices and sensors, alarms, and interlocks) and,*
>
> (vi)   *Pumps.*

There is no explicit reference to "critical equipment".

PROPOSED UPDATE

> *Expanding paragraph (j) to cover the mechanical integrity of any critical equipment.*

DISCUSSION

At first, the proposed update sounds perfectly sensible. Those equipment and instrument items that are deemed critical to process safety should receive priority attention. But, as we saw in Chapter 12, determining which equipment and instrument items are critical is not always as easy as it sounds. In the example provided in that chapter the operating technicians may frequently refer to a local pressure gauge when checking on the status of a reactor. Yet an outsider may not be aware of the reliance that is placed on that gauge, and so deem it to be non-critical. As with so many other elements of process safety, there is built-in subjective and circularity, a topic that we discuss in The Process Safety Roundabout (Sutton, The Process Safety Roundabout, 2022).

CHAPTER 20

EQUIPMENT DEFICIENCIES

The need to correct equipment deficiencies is fundamental to a process safety management program. As with so many other aspects of process safety, the challenge lies in defining the limits. As what point does an equipment item become "deficient"?

The content of this chapter has links to the material provided in Chapter 12 — Definition of Critical Equipment, and Chapter 20 — Mechanical Integrity of Critical Equipment.

THE REGULATION

The topic of this chapter is to do with paragraph (j): Mechanical Integrity.

- (a) Application
- (b) Definitions
- (c) Employee Participation
- (d) Process Safety Information

(e)   Process Hazards Analysis

(f)   Operating Procedures

(g)   Training

(h)   Contractors

(i)   Prestartup Safety Review

(j)   ***Mechanical Integrity***

(k)   Hot Work

(l)   Management of Change

(m)   Incident Investigation

(n)   Emergency Planning and Response

(o)   Compliance Audits

(p)   Trade Secrets

Subparagraph (5) of paragraph (j) to do with equipment deficiencies reads as follows,

> *The employer shall correct deficiencies in equipment that are outside acceptable limits (defined by the process safety information in paragraph (d) of this section) before further use or in a safe and timely manner when necessary means are taken to assure safe operation.*

PROPOSED UPDATE

> *Clarifying paragraph (j) to better explain "equipment deficiencies"*

"Acceptable limits" are often hard to define. The layers of process safety information shown in Figure 20.1 provide guidance.

DISCUSSION

OSHA has not provided guidance as to what they are looking for with regard to tightening the definition of equipment deficiencies. It is likely that their concern stems from the fact that many incidents, including those reported on by the Chemical Safety Board, were caused — at least in part — by some type of equipment deficiency.

In the current standard OSHA has defined equipment deficiency in terms of the acceptable limits that are integral the process safety information. Paragraphs (d)(2)(C) and (D) are to do with the technology of the process. They require that the following information be provided.

> (C) Maximum intended inventory;
> (D) Safe upper and lower limits for such items as temperatures, pressures, flows or compositions;

Defining safe limits is not easy, as can be seen in Figure 20.1 (Sutton, Process Risk and Reliability Management, 2014). The value shown could be for any operating parameter, but will usually be flow, level, temperature and pressure.

Figure 20.1: Operating, Safe and Emergency Limits

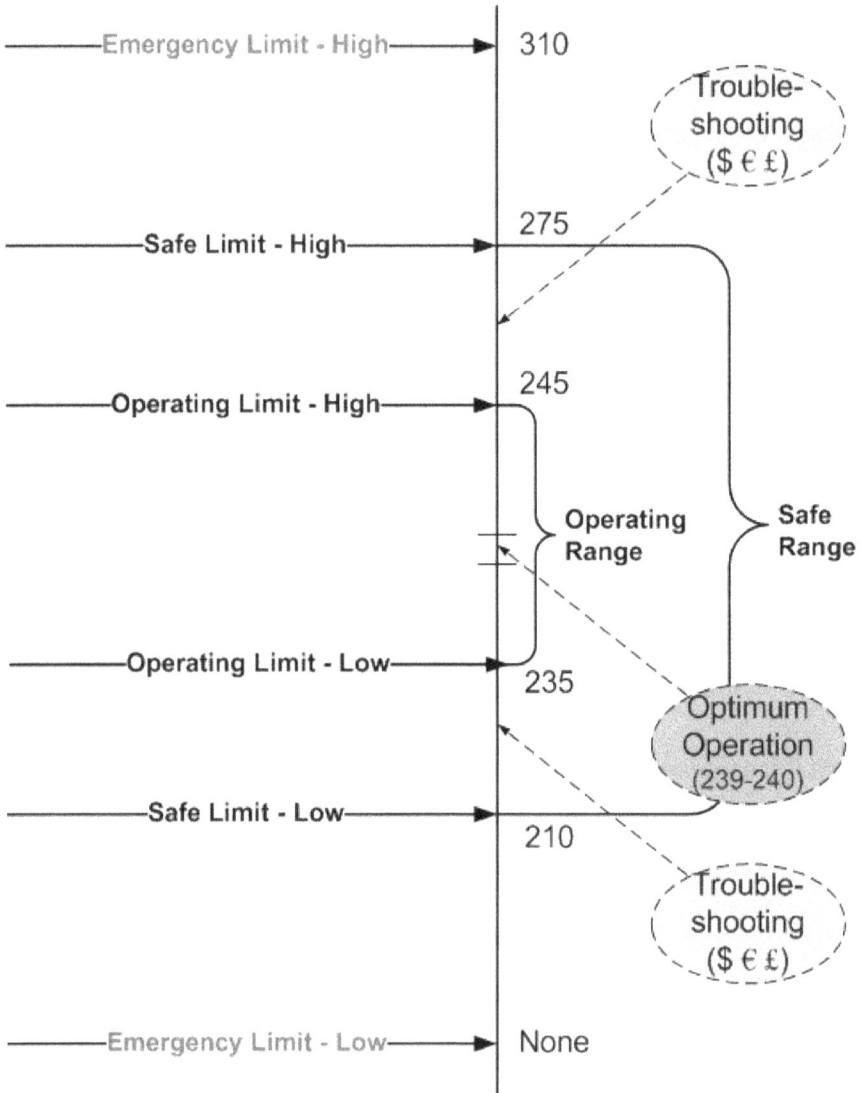

If operating conditions are allowed to move outside the operating limits, but within the safe limits, then the facility is said to be in trouble, *i.e.*, there are no safety issues to worry about, but the system is operating inefficiently. Troubleshooting efforts to bring the value back into the operating range will save money. Indeed, much of management's attention will be directed toward trouble shooting because addressing difficulties in this area will often lead to a significant improvement in profitability for relatively little expenditure. Examples of 'trouble' include:

- Excessive energy consumption,
- Product quality problems,
- Unusually high consumption of spare parts, and
- Low production rates.

The operating limit values are often quite fuzzy. As the system moves away from optimum operation it will start to exhibit symptoms of unusual operation which will eventually lead into the troubleshooting range.

The next range is defined by the *safe limit* values. In the case of Figure 20.1, were the parameter allowed to exceed 275 or go below 210 then the system is in an unsafe condition and action must be taken to bring that value back into the safe range.

The final set of values is the *emergency limits*. If the process parameter goes beyond one of these limits, then an emergency situation has been created. Immediate action is required; generally, the safety instrumentation and safety equipment (such as pressure relief valves) will be activated. In Figure 20.1 the upper emergency limit is 310; there is no lower emergency limit.

CHAPTER 21

ORGANIZATIONAL CHANGES

Management of Change systems generally focus on equipment and instruments. However, an effective Management of Change program will also consider the impact of changes to "soft" items, such as organizational structure and personnel changes, even though they are generally more difficult to categorize and define in terms of their impact on system safety.

THE REGULATION

(a) Application
(b) Definitions
(c) Employee Participation
(d) Process Safety Information
(e) Process Hazards Analysis
(f) Operating Procedures
(g) Training
(h) Contractors
(i) Prestartup Safety Review

(j)   Mechanical Integrity

(k)   Hot Work

(l)   Management of Change

(m)   Incident Investigation

(n)   Emergency Planning and Response

(o)   Compliance Audits

(p)   Trade Secrets

Paragraph (l) — Management of Change (MOC) — starts as follows.

> *The employer shall establish and implement*
> *written procedures to manage changes (except*
> *for "replacements in kind") to process chemicals,*
> *technology, equipment, and procedures; and, changes*
> *to facilities that affect a covered process.*

The paragraph does not mention changes to organizations or management systems.

PROPOSED UPDATE

> *Clarifying that paragraph (l) covers organizational*
> *changes.*

DISCUSSION

Managing organizational change is challenging because it involves human behavior and feelings — issues which are difficult to understand and to predict. For example, a large theoretical gain in efficiency may be achievable if operators and maintenance personnel can share their work activities. Yet, such changes can generate concerns about job security and loss of seniority. The upshot may be that that overall efficiency and productivity may actually fall were that change to be implemented.

The following are examples of organizational change that could have an impact on a facility's safety performance.

- A corporate directive calls for a reduction in the number of people employed at the site.
- The company comes under new ownership.
- Six of the most senior technicians retire within a three-month period — taking their knowledge and expertise with them.
- The operations superintendent proposes to change the route that delivery trucks follow within the facility boundaries.
- Shift workers vote for a change from eight-hour to twelve-hour shifts.
- The engineering manager suggests that a different contract company be used to bench-test the facility's relief valves.
- The Information Technology department installs a new computer system for inventory control.
- A purchasing agent decides to use a different vendor to supply a critical spare part.
- There is a general reduction in head count because the company is in financial difficulties.

Personnel and organizational changes take place constantly. The challenge lies in knowing where to draw the line. In the case of the first example — the retirement of the senior technicians — there may be no need for an MOC review if the training program for new employees is sufficiently thorough. However, a significant change in head count probably should be carefully evaluated.

CHAPTER 22

ROOT CAUSE ANALYSIS

An 800 person forum comprised of Root Cause Analysis (RCA) practitioners from all over the world tried to define "Root Cause Analysis." They could not agree on an answer. . . It means different things to different industries – even different things within the same industries. It is even difficult to find consistency within the same companies, or even sites within a company.

(Nelms, 2007)

A fundamental feature of process safety management is that it is a systems discipline. The management elements discussed in previous chapters are rarely handled in isolation — they interact with one another in ways that are often difficult to understand, or even identify.

This complexity is evident when conducting an incident investigation following a serious incident or near miss. It is very unusual for such an incident to have a single cause. There is usually a string of events. Each of these events has one or more causes; those causes are themselves events, which have causes, and so on and so on. For example, the incident may be "Tank Overflow". Analysis of this incident could lead to the following chain of questions and answers.

- Why did the tank overflow?
  Because the level control instrumentation failed.
- Why did the instrumentation fail?
  Because it had not been properly maintained.
- Why was it not properly maintained?
  Because the maintenance documentation was not available.
- Why was the documentation not available?
  Because the instrument supplier changed ownership.
- Why ... ?

Even this simple example may miss some lines of investigation. For example, in response to the question "Why did the instrumentation fail?" other answers could be:

- Spare parts had not arrived in time.
- The instrument technicians had not been properly trained.
- The chemical composition of the contents of the tank had changed.
- ...

In situations such as these, it is helpful to conduct a root cause analysis.

The value of root cause analysis is that its conclusions can address a much broader range of issues than those immediately to do with the event being investigated. In the case of the tank overflow example, the analysis may show that there are systemic problems to do with training, or with the company's procurement system, or with communications with equipment vendors.

There is, however, a problem with root cause analysis — there are, in fact, no true *root* causes. There are events which have causes; those causes are themselves are events, which have their own causes. The string can regress indefinitely, thus creating what has been referred to as the 'Root Cause Myth' (Gano, 2007). It is possible to find root causes, but not "the" root cause.

In practice, a root cause tends to be one that can be addressed within the realities of the current management system and operating constraints.

THE REGULATION

(a)   Application.
(b)   Definitions
(c)   Employee Participation
(d)   Process Safety Information
(e)   Process Hazards Analysis
(f)   Operating Procedures
(g)   Training
(h)   Contractors
(i)   Prestartup Safety Review
(j)   Mechanical Integrity
(k)   Hot Work
(l)   Management of Change
(m)   ***Incident Investigation***
(n)   Emergency Planning and Response
(o)   Compliance Audits
(p)   Trade Secrets

Paragraph (m) — Incident Investigation — provides instructions as to what constitutes an incident, and how it is to be investigated.

However, the only part of the regulation that discusses potential root causes is (m)(4)(iv) — "the factors that contributed to the incident".

PROPOSED UPDATE

The proposed update reads as follows,

> *Amending paragraph (m) to require root cause analysis*

Given the just discussed difficulties with defining the term root caused analysis, this update requires explanation and clarification.

OSHA / EPA FACT SHEET

In their Fact Sheet, *The Importance of Root Cause Analysis During Incident Investigation*, (Occupational Safety and Health Administration, 2016) OSHA and the EPA say,

> *A successful root cause analysis identifies all root causes—there are often more than one.*

This statement is disingenous — there is *always* more than one cause of an incident, but none of them are a true "root cause".

Additional information is provided in the ebook Incident Investigation and Root Cause Analysis (Sutton, Incident Investigation and Root Cause Analysis, 2022).

DISCUSSION

As we have seen, root cause analysis is a challenging exercise. In addition to the difficulties just described, there are many other issues to consider, some of which are described below.

FIXATION

During discussions people tend to take up a particular point of view, and then obstinately defend it, even when they are proven to be wrong. In other words, they become fixated. These people develop pride of ownership in their opinions. In other words, people — and that includes all of us at one time or another — tend to pick on one or two events, or perceived causes of events, and will not change their mind from that point forward. In the literal sense of the word, they are *prejudiced,* because they pre-judge events and the causes of events.

SINGLE/MULTIPLE INCIDENTS

Care has to be taken when developing root causes from the analysis of just one single incident. For example, a single investigation may note that a cause of an event being examined was that the operating procedure was not correct. However, the team is not justified, based on this one incident, in saying that a root cause is 'problems with operating procedures'. However, if the same facility conducts say 24 additional investigations and finds that operating procedures are often a factor, then it does indeed appear that problems with procedures are indeed a root cause.

Only when a facility has conducted a large number of incident analyses is it legitimate to talk about true root causes. In one facility, it became apparent, after 25 investigations (including near misses) that the following three issues kept bubbling up.

- Difficulties to do with communications at the operating company/contractor/subcontractor interfaces.

- Failure to recognize that some apparently routine work was, in fact, creating changes that should have been analyzed through the Management of Change system.

- Lack of knowledge of the availability of technical information.

LEVELS OF ROOT CAUSE

One of the difficulties associated with root cause analysis lies in determining the level at which to stop. After all, if the causes of an incident are pursued for long enough the team will eventually be discussing the philosophical, moral and theological issues to do with human nature. This is obviously absurd; a sensible stopping point is required.

CHAPTER 23

LOCAL EMERGENCY RESPONSE

The goal of process safety programs is to prevent serious incidents from occurring. Yet, no matter how good the program may be there is always the possibility of a serious chemical release or fire. At those times an immediate response is required.

THE REGULATION

(a)	Application.
(b)	Definitions
(c)	Employee Participation
(d)	Process Safety Information
(e)	Process Hazards Analysis
(f)	Operating Procedures
(g)	Training
(h)	Contractors
(i)	Prestartup Safety Review
(j)	Mechanical Integrity
(k)	Hot Work

(l)   Management of Change
(m)   Incident Investigation
(n)   ***Emergency Planning and Response***
(o)   Compliance Audits
(p)   Trade Secrets

The regulation currently reads as follows,

> *The employer shall establish and implement an emergency action plan for the entire plant in accordance with the provisions of 29 CFR 1910.38. In addition, the emergency action plan shall include procedures for handling small releases. Employers covered under this standard may also be subject to the hazardous waste and emergency response provisions contained in 29 CFR 1910.120 (a), (p) and (q).*

PROPOSED UPDATE

> *Revising paragraph (n) to require coordination of emergency planning with local emergency-response authorities.*

With regard to emergency response, OSHA refers to 29 CFR 1910.38 — Emergency action plans (Occupational Safety and Health Administration, 2002). This standard consists of six top level paragraphs. None of them refer explicitly working with local emergency planning with local authorities, although it would make sense to do so.

This proposed change may also involve coordination with the EPA's Risk Management Program actions to do with emergency response.

DISCUSSION

Guidance to do with emergency planning is provided in the ebook *Emergency Management* (Sutton, Emergency Management, 2023).

LEVELS OF EMERGENCY

An emergency response can be divided into three phases, based on the steps shown in Figure 20.1.

EMERGENCY — PHASE ONE

The system is at or near the emergency limits, but the operators and supervisors believe that they are able to return the plant to normal conditions using normal operating procedures and techniques. It is critical that they understand the exact nature of the problem if they are to be successful in this. Many accidents would have been less severe had the operators not tried to "fight" the situation, but simply shut down the facility in an orderly manner. (On the other hand, a full facility shutdown is not always the best response to an incipient emergency because doing so increases the number of actions that the operator has to perform and it can stress many equipment items. The advantage of keeping the unit running is that the operators can concentrate on correcting the emergency situation. They do not have to simultaneously cope with bringing down all the other equipment in a safe manner. Moreover, the avoidance of a full shutdown means that the unit can be brought back on line relatively quickly with minimal production loss).

EMERGENCY — PHASE TWO

The second phase of an emergency occurs when the safety instrumented system and other high reliability, automated devices (including relief valves) take over. At this point in time the role of the operator is simply to secure the unit as it shuts down.

EMERGENCY — PHASE THREE

In the third phase of an emergency, the situation is out of control. There may be a large fire or chemical release to contend with. The full emergency response system is needed to minimize injuries, environmental damage and loss of equipment.

Figure 23.1 provides more detail to do with the third phase. It shows the ways in which emergencies can be initiated, along with the appropriate levels of response. It also shows were Local Emergency Response fits into the overall picture.

Figure 23.1. Levels of Emergency

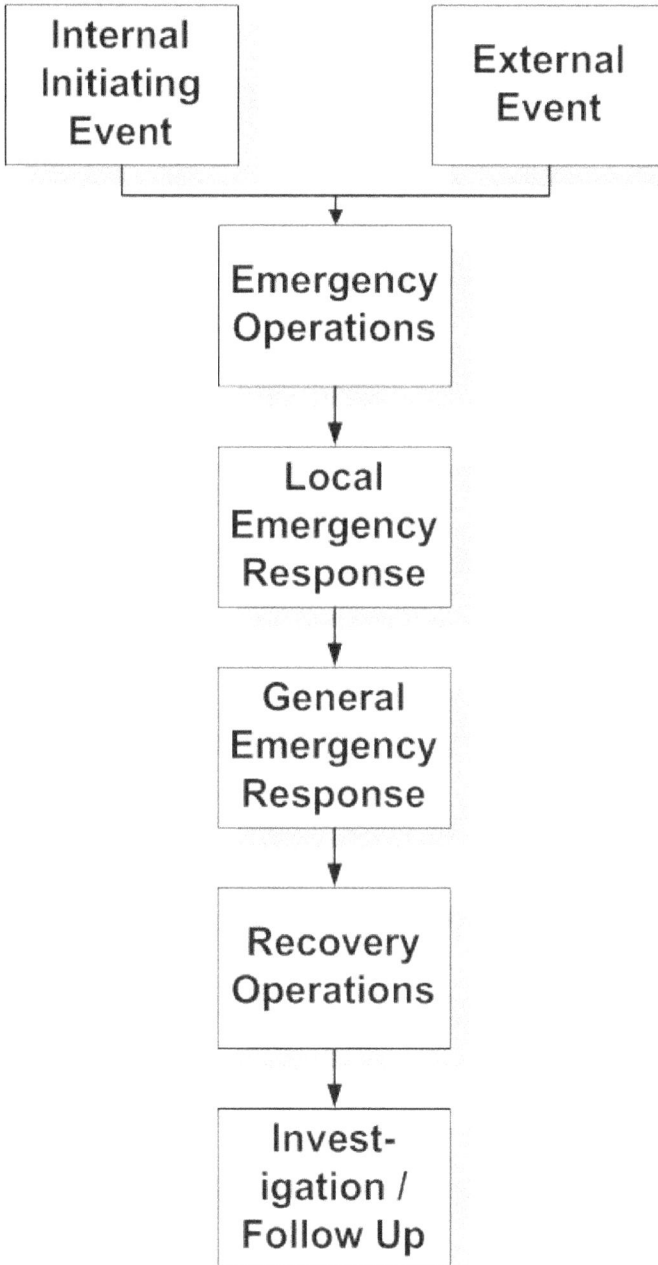

CHAPTER 24

THIRD-PARTY COMPLIANCE AUDITS

*There is always news about safety, and some of that
news will be bad.*

A udits are a fundamental component of all management programs. Process safety management is no different — there are always gaps between what "should be" and "what is". It is vital that management be informed of these gaps on a timely basis, and that they then take the appropriate action.

The purpose of an audit is to identify those gaps. It is *not* the purpose of an audit to recommend ways of the closing the identified gaps — doing so is follow up work to be carried out by management once the auditors have left the facility.

THE REGULATION

This proposed update to the standard is to do with paragraph (o) of the regulation.

(a)   Application.

(b)   Definitions

(c)   Employee Participation

(d)   Process Safety Information

(e)   Process Hazards Analysis

(f)   Operating Procedures

(g)   Training

(h)   Contractors

(i)   Prestartup Safety Review

(j)   Mechanical Integrity

(k)   Hot Work

(l)   Management of Change

(m)   Incident Investigation

(n)   Emergency Planning and Response

(o)   ***Compliance Audits***

(p)   Trade Secrets

The current regulatory audit requirements are,

(1) *Employers shall certify that they have evaluated compliance with the provisions of this section at least every three years to verify that the procedures and practices developed under the standard are adequate and are being followed.*

(2) *The compliance audit shall be conducted by at least one person knowledgeable in the process.*

(3) *A report of the findings of the audit shall be developed.*

(4) *The employer shall promptly determine and document an appropriate response to each of the findings of the compliance audit, and document that deficiencies have been corrected.*

(5) *Employers shall retain the two (2) most recent compliance audit reports.*

PROPOSED UPDATE

Amending paragraph (o) to require third-party compliance audits

DISCUSSION

Currently, OSHA does not require that companies use outside or third-party auditors. Presumably, they are introducing this measure because they are concerned that in-house auditors may be under pressure from management to minimize the seriousness of their findings.

THIRD-PARTY AUDITOR

OSHA has not defined what they mean by "third-party auditor", but a reasonable assumption is that such an auditor would come from the third group: an outside company that specializes in process safety work.

If a company does use an outside auditor, then that auditor is likely to come from one of the following four groups.

- A regulatory agency;
- Attorneys representing plaintiffs following an accident;
- An outside company that specializes in process safety work; or
- A person from within the client company, but based in another department.

The use of outside companies should help make the audit more objective. Outsiders are not likely to be familiar with the processes being audited, nor are they involved with the inter-personal relationships to o with those running the facility. For them, this project is only one of many, so they will not be hesitant about issuing critical findings.

A second benefit of using an outsider is that he or she will bring a fresh perspective based on their knowledge of how other companies and facilities operate. In particular, he or she may identify hazardous situations that employees at a facility have learned to live with but that would not be accepted as being safe elsewhere.

JUST AN AUDITOR

An outside auditor should not provide follow-up services to the audit. Such services represent a conflict of interest, with the exception of reviewer-type comments, as discussed above.

COMPLIANCE

Both the OSHA standard and the proposed update use the work "compliance". This word creates difficulties in the context of process safety management. Most safety standards are prescriptive, so it is straightforward for the auditor to determine compliance. For example, if a regulation requires that a maintenance worker requires a safe work permit before starting work, then the auditor can determine if that permit was actually completed. It is a simple "Yes/No" situation.

When it comes to non-prescriptive/performance-based regulations, an auditor is in a much more difficult situation. There are no "Yes/No" answers. The only way to be "in compliance" is not to have incidents. Yet no company, no matter how stellar its safety performance, can achieve that level of perfection. Therefore, by definition, all companies are not in compliance. This means that the auditor's assessment is going to be somewhat subjective.

Ultimately, as we have already seen, the only measure of success is success. If a facility does not have incidents, then it is "in compliance". On the other hand, the occurrence of an event is evidence of "non-compliance".

CHAPTER 25

REVIEW OF PSM MANAGEMENT SYSTEMS

O SHA's process safety regulation does not require companies to review their management systems. That they should do so is, of course, a sensible requirement. Indeed, all management systems need to be reviewed on a regular basis.

The audit process that was discussed in the previous chapter (paragraph (o) of the regulation) will not only identify gaps in performance, but it can help identify ways in which the management process itself could be improved. Nevertheless, a different process for evaluating and improving management systems is needed. Audits measure performance against an existing standard; management reviews evaluate the management systems that are being used to meet regulatory requirements.

When conducting a PSM review there are four questions to ask.

1. What is our current status?
2. Where are we most vulnerable?
3. How are we progressing?

4. How do we compare to others?

None of these questions fall under the purview of the normal audit process.

THE REGULATION

OSHA did not assign this update to any particular element. In this chapter it is placed in the Application category.

(a)   ***Application***
(b)   Definitions
(c)   Employee Participation
(d)   Process Safety Information
(e)   Process Hazards Analysis
(f)   Operating Procedures
(g)   Training
(h)   Contractors
(i)   Prestartup Safety Review
(j)   Mechanical Integrity
(k)   Hot Work
(l)   Management of Change
(m)   Incident Investigation
(n)   Emergency Planning and Response
(o)   Compliance Audits
(p)   Trade Secrets

PROPOSED UPDATE

Including requirements for employers to develop a system for periodic review of and necessary revisions to their PSM management systems (previously referred to as "Evaluation and Corrective Action").

This proposed update is actually calling for an *assessment* of the management systems to determine if they are fit for purpose.

Note: There is a grammatical problem with this proposed update. The word management is repeated, as in, "Process Safety Management management systems". (We see the same difficulty with the commonly-used term 'PIN Number').

DISCUSSION

Audits can help identify ways of improving management systems. But, as discussed in Chapter 24, that is not really the purpose of an audit. The purpose of an audit is to evaluate performance against an existing standard, not to find ways of changing or improving that standard, nor to find ways of changing or improving the standard. A different way of evaluating the process safety program is called for. There is a need for *assessments* as distinct from *audits* to determine if the process safety program is fit for purpose.

Assessments will not be as objective as formal audits; they will allow scope for the subjective opinions of those conducting the assessments, and of the facility's managers. There is no "Yes/No" response as there is with a formal audit. (The terms Verification and Validation are sometimes used to make the same distinction between audits and assessments. Verification is concerned with ensuring that a facility meets the letter of a regulation or standard; validation is more to do with whether a facility's process safety program meets the spirit of the same regulation or standard.)

Like an audit, an assessment is built around a series of questions that identify gaps in a facility's safety performance. However, there are fundamental differences.

- Instead of a "Yes/No" answer, a numerical range can be used for the response, ranging from inadequate all the way to excellent.
- Different assessors can provide different responses to the same question. There are no "right answers".
- Unlike an audit, there is no requirement to answer all the questions.

The distinction between audits and assessments was made clear at one facility at which the audit/review team conducted two activities in parallel. The first was an OSHA-style audit. When it came to the client's operator training program the *audit* noted that all regulatory requirements were being met fully — a training program was in place and was being implemented. However, an *assessment* provided by the same team stated that there were so many problems with training that the best strategy may be to start a brand-new training program.

Chapter 24 — Third-Party Compliance Audits — discussed the need for auditors who are external to the organization being audited. Such is not the case with assessments. Indeed, the facility's managers should be actively involved in the assessment; there is no concern about conflicts of interest.

Another feature of audits is that they are not action plans. Having identified an issue that needs to be corrected, the audit team then moves on. It is up to the facility's managers to make those corrections. Such is not the case with assessments — all members of the team are encouraged to develop ideas for improving the management systems and in finding cost-effective solutions to the problems identified. (In practice, many auditors do offer advice and opinions. However, it should be understood that, when they do so, they are crossing the boundary between audits and assessments.)

Using operating procedures as an example, an assessor could take a particular procedure to do with starting a compressor and ask questions such as the following.

- Is this procedure too long?
- Is it too short?
- Has information from the equipment manufacturer been included?
- Can it be used outside at night in the pouring rain?
- Is it written at the correct grade level?
- And — the most important question of all — does anyone actually use this procedure, or does it sit on a shelf or a hard drive quietly gathering dust?

CHAPTER 26

WRITTEN PSM
PROCEDURES

One of the features of OSHA's process safety regulation is that it is does not require companies to develop written procedures for the overall program. Nor is there a formal reporting program (this is in contrast to the requirements of the EPA's Risk Management Program). This way of thinking is in line with the performance-based nature of the regulation. If a company avoids having serious events, then there is no need for further evidence that it is doing well. If written process safety procedures help a company achieve its safety goals, then the company managers should write such procedures. However, if that company can achieve its goals with minimal written procedures, then there is no need for further discussion. Whether it has a well-written program should be inconsequential.

This proposed update changes that policy — it is no longer sufficient for companies to do a good job when it comes to safety — they must also files reports to show that they are doing a good job.

THE REGULATION

There is no general requirement in the current standard for written procedures for all the management elements. (There are, however, some specific requirements in this area, for example to do with the updating of operating procedures.) Since OSHA did not assign this update to any particular element it has been placed in the Application category.

(a)   *Application*
(b)   Definitions
(c)   Employee Participation
(d)   Process Safety Information
(e)   Process Hazards Analysis
(f)   Operating Procedures
(g)   Training
(h)   Contractors
(i)   Prestartup Safety Review
(j)   Mechanical Integrity
(k)   Hot Work
(l)   Management of Change
(m)   Incident Investigation
(n)   Emergency Planning and Response
(o)   Compliance Audits
(p)   Trade Secrets

PROPOSED UPDATE

OSHA's proposed change reads as follows,

> *Requiring the development of written procedures for all elements specified in the standard, and to identify*

records required by the standard along with a records retention policy (previously referred to as "Written PSM Management Systems")

This update has two separate sections: (1) a requirement to develop written procedures for the process safety program, and (2) a requirement for records retention.

OSHA does not provide specific guidance to do with either of these proposed updates. Some general considerations are discussed below.

DISCUSSION

In practice, most companies already have a Process Safety Manual. Therefore, addressing the requirements of this update will probably create additional work, but the foundation will usually be already present.

DEFINITION OF GOALS

The written procedures should include a description of the goals of each of the management elements. This description may include an example that can be used as a "go-by" that can illustrate the goals that management has provided.

MANAGEMENT STRUCTURE

The Manual will need to provide a description of the overall management structure. Figure 26.1, which is taken from the book Process Risk and Reliability Management (Sutton, Process Risk and Reliability Management, 2014), provides a starting point.

Figure 26.1. Process Safety Organization Chart

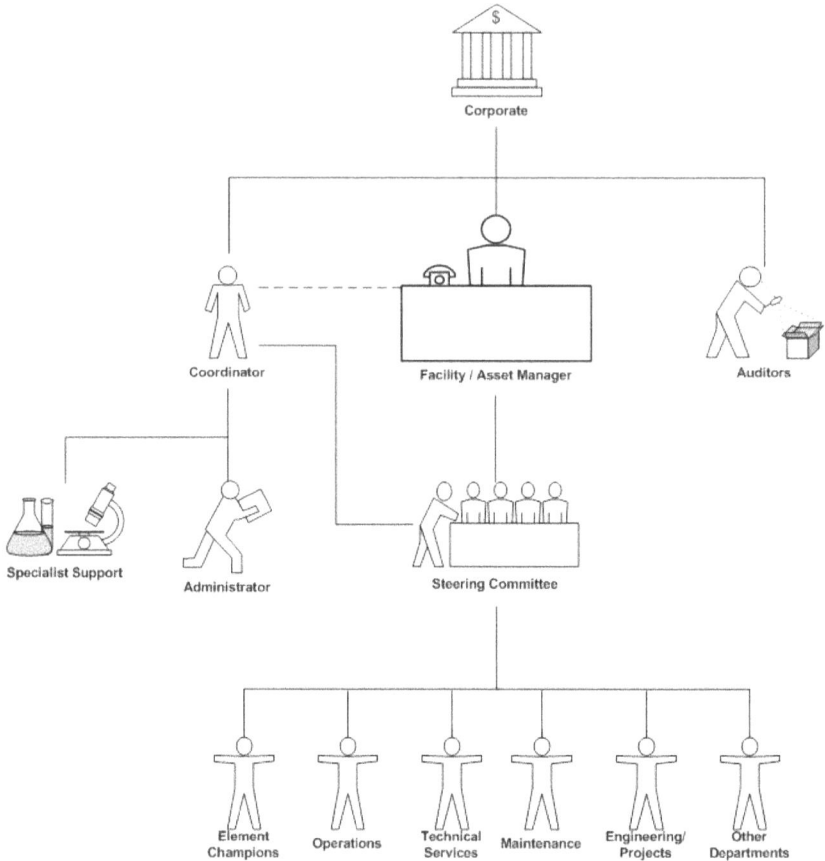

The sketch shows some of the key management functions. They include:

- Corporate direction,
- The facility or asset manager,
- The PSM coordinator,
- A steering committee,
- An administrator,
- PSM elements leads,
- Specialist support, and
- Auditors.

This part of the manual will also contain information to do with metrics and the baseline for each of the management elements.

OPERATING BINDERS

One way of physically organizing the risk management program is to create a library of binders (either electronic or on paper): one for each element. These binders will contain information associated with that element and each will comprise a section of a master binder. They will also contain indexing information that tells the user where other information can be found. As far as possible, the binders and their contents should look the same as one another, and should have similar Tables of Contents, following a layout such as the following.

- Introduction
- Objectives of the Program
- Regulations
- Industry Standards
- Company Standards
- Protocols
- Employee Participation
- Process Safety Information
- Administration
- Equipment Items Covered
- Personnel
- Use of Outside Companies
- Project Management
- Phases of the Program
- Budget
- Schedule

All the binders should be complete at all times. Initially, there will be very little detail in them, but they should nevertheless have a section for each part of each element of the standard. When printed, the binders should also be physically attractive and neatly organized in order to make them easy to use, and in order to make a good impression if the program is audited.

UPDATE POLICY

The Manual will include procedures for updating its contents as the organization changes. This function will usually be part of the overall management system software.

RECORDS RETENTION

A policy is needed to identify which records and procedures should be kept on file. They may be needed for a future audit or incident investigation.

CHAPTER 27

THEMES

The previous chapters have discussed OSHA's 24 proposed updates to its process safety management standard. From these specific updates it is possible to develop various themes and generalizations. In this chapter we will discuss some of these themes. We will also consider areas which could have been considered for update, but which OSHA did not include. The themes include:

- Definitions,
- Hazards Analysis,
- Systems Analysis,
- Operational Excellence,
- Culture, and
- Imagination.

DEFINITIONS

As we have seen, OSHA is defining terms such 'RAGAGEP' and 'critical equipment' more tightly. However, it worth considering factors that lead to potential confusion. These factors include understanding process safety limits and difficulties created by circular reasoning.

PROCESS SAFETY LIMITS

Many of the proposed changes are to do with defining process safety boundaries — not only for equipment, but also the interfaces with other organizations. As the refinery operator anecdote in Chapter 10 illustrates, there are no true, unbreachable physical boundaries. The element in Chapter 10 — Defining the Limits of a PSM-Covered Process — explicitly addresses this issue. But the definition of limits crops up implicitly in other elements such as the 'Definition of Critical Equipment' and 'Atmospheric Storage Tanks'. The discussion associated with Figure 20.1 provides a basis for determining limits. They include operating, trouble-shooting and emergency criteria.

CIRCULAR REASONING

A difficulty with many aspects of process safety management is to do with circular reasoning — a situation where a reasoner ends with a result that takes him back to the starting point. The colloquial term Catch-22 is to do with the same conundrum: the reasoner cannot escape a logical loop due to contradictory rules or assumptions.

The following are examples of circular reasoning in the context of process safety management.

Process Hazards Analysis
- Could high temperature cause an accident?
- What is high temperature?
- It is the temperature that could cause an accident.

RAGAGEP
- What is "good engineering practice"?

- It is that practice that is defined by engineering experts.
- What makes them recognized experts?
- They know how to recognize good engineering practice.

Management of Change
- Does the proposed change require that a formal Management of Change (MOC) analysis be carried out?
- We need to conduct an MOC to find out.

Critical Equipment
- Is this piece of equipment safety critical?
- What is meant by safety critical?
- The equipment item is considered to be critical to safety.

Risk Analysis
- Is a risk analysis needed?
- It is needed if the level of risk is unacceptable.
- How do we know if the risk is unacceptable?
- Conduct a risk analysis.

PROCESS HAZARDS ANALYSIS

Many of the proposed changes relate to paragraph (d) of the standard — Process Hazards Analysis (PHA). The role of a PHA team is to identify and risk rank hazards, not to develop solutions. This strategy makes sense; the members of a PHA team are not necessarily qualified to develop solutions. The team is not a one-minute engineering department. However, the 15th item in the list, which calls on PHA teams to consider safer technology and alternatives, may be changing this strategy.

HUMAN FACTORS AND RELIABILITY

Since the process safety standard was introduced 30 years ago, the topics of human factors, human reliability and human error have become prominent. The discussions have covered a range of topics such as,

- Protecting workers in the event of fire or chemical release. This topic can include PPE (Personal Protective Equipment), the siting of control rooms, and automating inherently hazardous tasks.
- Determining human error rates (Sutton, Safety Moment #103: Human Error Modeling, 2020). There are many aspects to this topic. For example, there is a significant distinction between slips and mistakes. The proposed updates do not provide guidance on issues such as these.
- Understanding that people respond differently in an emergency situation than they do during normal operations.

AUTOMATION

One of the most important changes in industrial practice since OSHA introduced its PSM standard is to do with the impact of electronics, in all of its many forms. The process industries have been no exception. Increasingly, automated systems and robotics are replacing human workers

This transformation affects safety positively in at least two ways. First, by replacing humans with automated systems and robots, the inherent safety of a facility is improved because, "If a man's not there, he can't be killed." Second, during an emergency, instruments — particularly Safety Instrumented Systems — will reliably take the correct actions needed to bring a facility to a safe state. Humans, on the other hand, are not at their best during an emergency.

There are references in the proposed updates to the "definition of critical equipment", "requiring safer technology and alternatives" and the "mechanical integrity of critical equipment". However, there is nothing in the updates to the standard that explicitly addresses the topic of automation, in spite of the benefits that it offers.

SUPPLY CHAIN MANAGEMENT

Advances in computer and software technology have created profound changes in the manner in which industries are organized and managed. The process and energy industries are no exception. Supply chains are much more tightly managed than they were 30 years ago, inventories have been reduced, as have spare parts and equipment items, and the number of employees. These changes have led to increased profitability, but they may have eaten into safety margins. Just-in-Time management systems have made facilities more vulnerable to interruptions in supply chains — there is less resilience in the system.

This increased vulnerability was made evident once commerce started to open up following the COVID pandemic peak. Indeed, the supply chains are still not operating as well as they were before the pandemic. There appears to be no evidence yet that a serious process-related incident can be attributed to COVID-related supply chain difficulties (that would be an intriguing root cause). Nevertheless, there has to be a concern that stresses in the supply chains may lead to a serious incident.

Difficulties with the reliability of supplies are likely to be exacerbated in coming years as material shortages become more severe due to resource depletion, and the apparent deterioration of the global economic system that has provided a *raison d'être* for the Just-in-Time philosophy. Nevertheless, the coming years are likely to see increased tension between the process safety management professionals and those who want to increase profitability by tightening the supply chains even further. Eventually, something is going to snap.

SITING

The topic of siting is included in the process hazards part of the regulation. In the 30 years since the standard was published a number of serious incidents have involved problems with siting, or the related topic of equipment layout. Yet OSHA does not mention either of these topics in their proposed updates.

SYSTEMS ANALYSIS

Process safety management is to do with the management of systems. These systems are complex, and they are very difficult to understand and manage because they include interactions between equipment, human behavior and organizational structure. The systems nature of process safety is something that OSHA discussed in the original regulation. However, the agency has not developed this line of thinking in the proposed updates. Each of the 24 items tend to be treated discretely; they are managed separately.

OPERATIONAL EXCELLENCE

In spite of its title, Process Safety Management is not fundamentally a safety program. It is a process for managing process facilities such that all aspects of their performance improve. Safety, production and productivity all get better. In other words, process safety management contributes to both Operational Integrity and Operational Excellence.

Figure 27.1 shows how the discipline of Process Safety Management contributes toward the larger topic of Operational Integrity. Topics such as RAM (Reliability, Availability and Maintainability), HSE (Health, Safety and Environment), Statistical Process Control, ISO 9000 and Occupational Safety each adds its own input to the overall mix. A facility which has a high level of operational integrity is one that performs as expected in an atmosphere of 'no surprises'. The facility exhibits integrity in all aspects of its operation.

Figure 27.1. Operational Integrity Management Programs

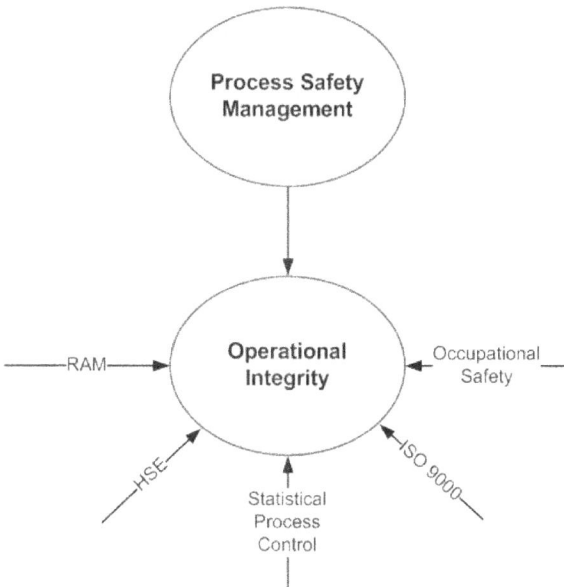

Operational Integrity can itself be seen as part of an even larger topic: Operational Excellence, as shown in Figure 27.2 (Sutton, A Brief History Of Process Safety Management, 2021). Whereas Operational Integrity is mostly made up of technical initiatives; operational excellence incorporates non-technical management systems that can affect safety and operability. These include distribution, inventory management, outsourcing, supply chain management and procurement.

Figure 27.2. Operational Excellence

CULTURE

In Chapter 13 it was stated that Employee Participation lies at the heart of any successful process safety program. That line of thought could be pursued further. After all, employee participation, along with the other elements of process safety, are fundamentally an expression of a company's culture. For this reason, other process safety systems have added the element 'culture' to their program. OSHA has not done so with these updates. This omission is somewhat surprising given that there has been considerable discussion in recent year to do with the topic of 'culture' within the context of process safety. For example, the Baker Commission report to do with BP's 2005 accident at Texas City (Baker, 2007) uses the word culture many times; the following is a quotation from that report,

> *BP has not instilled a common, unifying process safety culture among its U.S. refineries. Each refinery has its own separate and distinct process safety culture.*

DEFINING CULTURE

The Baker report does not define the word culture. However, the Center for Chemical Process Safety (CCPS) provides the following description of culture in a process safety context.

> *Process safety culture has been defined as, "the combination of group values and behaviors that determine the manner in which process safety is managed". More succinct definitions include, "How we do things around here," "What we expect here," and "How we behave when no one is watching." In an especially sound culture, deeply held values are reflected in the group's actions, and newcomers are expected to endorse these values in order to remain part of the group.*

Arendt divides the topic of culture into twelve elements (Arendt, 2009):

1. Establish safety as a core value.
2. Provide strong leadership.
3. Establish and enforce high standards of performance.
4. Formalize the safety culture emphasis/approach.
5. Maintain a sense of vulnerability.
6. Empower individual to successfully fulfill their safety responsibilities.
7. Defer to expertise.
8. Ensure open and effective communications.
9. Establish a questioning/learning environment.
10. Foster mutual trust.
11. Provide timely response to safety issues and concerns.
12. Provide continuous monitoring of performance.

A difficulty with definitions such as these is that it is not easy to come up with objective, quantified measures for them.

THE BIG CREW CHANGE

One practical aspect of maintaining a safe culture is to do with the loss of experienced people from a company or a facility. This phenomenon is sometimes referred to as the 'Big Crew Change'.

The loss of key personnel has been a particularly serious problem in the energy business in recent years — as the price of oil declined, more and more people were let go or "took the package". Then, when business picked up, those people were no longer available. Moreover, the experienced people, because they generally earned the highest salaries — were often the first to leave.

When these people leave the organization, they take their knowledge and experience with them. This is why on-going training programs are fundamental to safety — it is vital that people with less experience are able to achieve the same high level of performance once their predecessors have moved on.

IMAGINATION

In the year 2022 the Institution of Chemical Engineers held its annual process safety conference — Hazards32 — in Harrogate, England (Sutton, Net Zero by 2050, 2022). Many speakers and attendees, both in their presentations and in informal conversations, suggested that, after 30 years, process safety management was mature. This is both a compliment and a cause for concern. It is a compliment because there are fewer serious incidents now than there were when the discipline of process safety was being developed. The hard work that has been put in by so many people over the last 30 years has, to a considerable degree, paid off. (However, the incidents described by the Chemical Safety Board in previous chapters demonstrate that there are still many issues that need to be resolved.) The concern is to do with the fact that there does not seem to be much innovation or fresh thinking within the discipline. There is a danger that process safety professionals could become complacent. In other words, there is a lack of imagination.

The same concern applies to the changes discussed in the previous chapters. Most of the changes are quite specific — they address some of the limitations, gaps and ambiguities that exist in the current standard. There is little in what OSHA has presented that calls on companies to manage process safety in new or innovative ways (although the sections to do with 'Safer Technology and Alternatives Analysis' and 'Natural Disasters and Extreme Temperatures' are somewhat open-ended). The proposed updates to the standard do not consider how a company's process safety program could improve all aspects of the operation, not just safety, as discussed in the section to do with Operational Excellence.

After 30 years it is rather disappointing that OSHA has not proposed more fundamental changes to its process safety standard. The world of process safety has changed; the standards and regulations need to reflect some of this new and different world.

GLOSSARY

Abbreviation	Meaning
BSEE	Bureau of Safety and Environmental Enforcement (United States)
CAS	Chemical Abstract Service
CCPS	Center for Chemical Process Safety
CSB	Chemical Safety Board
DWH	Deepwater Horizon (drilling rig at the Macondo prospect)
EPA	Environmental Protection Agency (United States)
HSE	Health, Safety and Environmental
OSHA	Occupational Safety and Health Administration (United States)
NAICS	North American Industry Classification System
PHA	Process Hazards Analysis
PSM	Process Safety Management
RAM	Reliability, Availability and Maintainability
RMP	Risk Management Program (EPA)
SEMS	Safety and Environmental System (United States)

WORKS CITED

Arendt, S. (2009, April 22). Evaluate HSE/Process Safety Culture. *LAI/MEP Forum, Lake Charles, LA.*

Baker, J. (2007). *The Report of the U.S. Refineries Independent Safety Review Panel.*

Bureau of Safety and Environmental Enforcement. (2022). *Safety and Environmental Management Systems - SEMS.* Retrieved from https://www.bsee.gov/reporting-and-prevention/safety-and-environmental-management-systems

Chemical Safety Board. (n.d.). Retrieved from Reactive Hazards: https://www.csb.gov/reactive-hazards/

Chemical Safety Board. (2018, April). *CSB's Drivers of Critical Chemical Safety Change.* Retrieved from https://www.csb.gov/assets/1/6/csb_cdl_fact_sheet_-_psm.pdf

Chemical Safety Board. (2019, June 12). *Pryor Trust Fatal Gas Well Blowout and Fire.* Retrieved from https://www.csb.gov/pryor-trust-fatal-gas-well-blowout-and-fire/

Chemical Safety Board. (2021, July 29). Retrieved from https://www.youtube.com/watch?v=jh2HWT8gPeY

Cullen, D. (1990). *The Public Inquiry into the Piper Alpha Disaster.* London: Department of Energy, HMSO Cm 1310.

Environmental Protection Agency. (2022, August 31). *Accidental Release Prevention Requirements.* Retrieved from https://www.govinfo.gov/content/pkg/FR-2022-08-31/pdf/2022-18249.pdf

Environmental Protection Agency. (2022, July 12). *Complying with process safety information (PSI) resulting from new and updated recognized and generally accepted good engineering practices (RAGAGEP)*. Retrieved from https://www.epa.gov/rmp/complying-process-safety-information-psi-resulting-new-and-updated-recognized-and-generally

Environmental Protection Agency. (2022, November 1). *Risk Management Program (RMP) Rule*. Retrieved from https://www.epa.gov/rmp

Gano, D. (2007). *Gano, Dean*. Atlas Books.

HSI. (2022). *OSHA Updates PSM Standards for Retail Exemption*. Retrieved from https://hsi.com/blog/osha-updates-psm-standards-for-retail-exemption

Nelms, C. R. (2007). The Problem with Root Cause Analysis.

Occupational Health and Safety Administration. (n.d.). *Chemical Hazards and Toxic Substances*. Retrieved from https://www.osha.gov/chemical-hazards

Occupational Safety and Health Administration. (1992). Retrieved from 1910.119 – Process safety management of highly hazardous chemicals: https://www.osha.gov/laws-regs/regulations/standardnumber/1910/1910.119

Occupational Safety and Health Administration. (1997, February 28). *Standard Interpretations Akzo-Nobel Chemicals - Limits of a Process*. Retrieved from https://www.osha.gov/laws-regs/standardinterpretations/1997-02-28

Occupational Safety and Health Administration. (1999, December 20). *OSHA Archive*. Retrieved from https://www.osha.gov/laws-regs/standardinterpretations/1999-12-20

Occupational Safety and Health Administration. (2000). *Process Safety Management. OSHA 3132*. Retrieved from https://www.osha.gov/sites/default/files/publications/osha3132.pdf

Occupational Safety and Health Administration. (2002, November 7). *Emergency action plans*. Retrieved from https://www.osha.gov/laws-regs/regulations/standardnumber/1910/1910.38

Occupational Safety and Health Administration. (2016, July 18). *Covered Concentrations of Listed Appendix A Chemicals*.

Retrieved from https://www.osha.gov/laws-regs/standardin-terpretations/2016-07-21

Occupational Safety and Health Administration. (2016, May 11). *RAGAGEP in Process Safety Management Enforcement.* Retrieved from https://www.osha.gov/laws-regs/standardin-terpretations/2016-05-11-0

Occupational Safety and Health Administration. (2016, October). *The Importance of Root Cause Analysis During Incident Investigation.* Retrieved from https://www.osha.gov/sites/default/files/publications/OSHA3895.pdf

Occupational Safety and Health Administration. (2017). *Process Safety Management for Explosives and Pyrotechnics Manufacturing.* Retrieved from https://www.osha.gov/sites/default/files/publications/OSHA3912.pdf

Occupational Safety and Health Administration. (2017). *Process Safety Management for Storage Facilities.* Retrieved from https://www.osha.gov/sites/default/files/publications/OSHA3912.pdf

Occupational Safety and Health Administration. (2018, April 30). *Process Safety Management Retail Exemption Enforcement Policy.* Retrieved from https://www.osha.gov/laws-regs/standardinterpretations/2018-04-30

Occupational Safety and Health Administration. (2022). *Docket no. OSHA–2013–0020.* Retrieved from https://www.regulations.gov

Occupational Safety and Health Administration. (2022, August 30). *Process Safety Management (PSM Stakeholder Meeting.* Retrieved from https://www.govinfo.gov/content/pkg/FR-2022-08-30/pdf/2022-18614.pdf

Occupational Safety and Health Administration. (n.d.). *Chemical Reactivity Hazards.* Retrieved from https://www.osha.gov/chemical-reactivity

Occupational Safety and Health Administration. (n.d.). *Chemical*

Reactivity Hazards. Retrieved from https://www.osha.gov/ chemical-reactivity/standards

OSHA. (1992). *1910.119 App A - List of Highly Hazardous Chemicals, Toxics and Reactives (Mandatory)*. Retrieved from Occupational Safety and Health Administration: https://www.osha.gov/ laws-regs/regulations/standardnumber/1910/1910.119AppA

Sutton, I. (2014). *Process Risk and Reliability Management*. Elsevier.

Sutton, I. (2017). *Plant Design and Operations*. Elsevier.

Sutton, I. (2017). *Sutton Technical Books*. Retrieved from Asset Integrity: https://iansutton.com/ebooks/asset-integrity

Sutton, I. (2018). *Frequency Analysis*. Retrieved from Sutton Technical Books: https://iansutton.com/ebooks/frequency-analysis

Sutton, I. (2020). *Safety Moment #103: Human Error Modeling*. Retrieved from Sutton Technical Books: https://iansutton. com/index.php/safety-moments/safety-moment-103-human-error-modeling

Sutton, I. (2020). *Safety Moment #94: Fault Tree Analysis*. Retrieved from Sutton Technical Books: https://iansutton.com/index. php/safety-moments/safety-moment-94-fault-tree-analysis

Sutton, I. (2021). *A Brief History Of Process Safety Management*. Retrieved from https://iansutton.com/ebooks/brief-history-process-safety-management

Sutton, I. (2022, September). Retrieved from Sutton Technical Books: https://iansutton.com/safety-moments/safety-moment-73-storage-tanks-process-energy-industries

Sutton, I. (2022). Retrieved from Net Zero by 2050: https://netzero2050.substack.com/

Sutton, I. (2022, October 6). *Comment Submitted to OSHA*. Retrieved from Net Zero by 2050: https://netzero2050.substack.com/p/comment-submitted-to-osha

Sutton, I. (2022). *Incident Investigation and Root Cause Analysis*. Retrieved from Sutton Technical Books: https://iansutton.com/

ebooks/incident-investigation-root-cause-analysis

Sutton, I. (2022). The Process Safety Professional in a Net Zero World. *Hazards32*. Institution of Chemical Engineers.

Sutton, I. (2022, November 14). *The Process Safety Roundabout*. Retrieved from Sutton Technical Books: https://netzero2050. substack.com/p/the-process-safety-roundabout

Sutton, I. (2022). *Update to OSHA's Process Safety Management Regulation: An Index*. Retrieved from Net Zero by 2050: https://netzero2050.substack.com/p/update-to-os-has-process-safety-management

Sutton, I. (2023, January). Retrieved from Emergency Management: https://iansutton.com/ebooks/emergency-management

Sutton, I. (2023). *Update to OSHA's Process Safety Management Regulation: An Index*. Retrieved from Net Zero by 2050: https://netzero2050.substack.com/p/update-to-os-has-process-safety-management

Sutton, I. (n.d.). *Offshore Safety Management*. Elsevier.

Sutton, I. (n.d.). *Safety Moment #64: Equipment Spacing*. Retrieved from Sutton Technical Books: https://iansutton.com/safe-ty-moments/safety-moment-64-equipment-spacing-hy-drocarbon-storage-tanks

INDEX

www.ingramcontent.com/pod-product-compliance
Lightning Source LLC
Chambersburg PA
CBHW020704270326
41928CB00005B/257